Creativity, Heritage and the City

Volume 10

Cities are faced with various problems, including terrorism, energy challenges, and environmental issues, as well as inter-urban competition brought about by expanding globalization forces. What is required is to gather theoretical insights from various scientific areas, not only social science–humanities but also natural science, and connect them to the practical insights already gained through numerous efforts to deal with these issues on the ground. In this way, paradigms for urban creativity can be developed and we can start to accrue dependable practice and theoretically based intelligence that can be used for improved policymaking. The keywords for this book series are "urban creativity", "(cultural) heritage", and "social development". Developing cultural and natural resources, including heritage, so as to take the lead in evaluating, implementing, and suggesting urban or regional designs that harmonize ecology, society, and people, and to further develop urban and regional culture is essential. There is a particular focus in this book series on fostering individuals who can design, manage, and direct models, technologies, and tools for promoting interfaces between such actors as policymakers, urban planners, engineers, and residents. The above-stated goals can be implemented through cooperation with international research communities and networks, international organizations, and natural history institutions, academies of science, and research institutes.

Baher I. Farahat · Helga O. Alaeddine

Urban Acupuncture for Place Attachment and Social Resilience in Heritage Contexts

Springer

Baher I. Farahat
Architecture—Design and Built Environment
Beirut Arab University
Beirut, Lebanon

Helga O. Alaeddine
Human Sciences
Beirut Arab University
Beirut, Lebanon

ISSN 2366-4584 ISSN 2366-4592 (electronic)
Creativity, Heritage and the City
ISBN 978-981-95-8969-2 ISBN 978-981-95-8970-8 (eBook)
https://doi.org/10.1007/978-981-95-8970-8

Cover illustration: Le Pont de Maincy (part) by Paul Cezanne, Photo © RMN-Grand Palais (Muse´e d'Orsay) / Herve´ Lewandowski

This Springer imprint is published by the registered company Springer Nature Singapore Pte Ltd.
The registered company address is: 152 Beach Road, #21-01/04 Gateway East, Singapore 189721, Singapore

Preface

Cities remember. They remember through the weathering of stone and plaster, through the quiet repetition of everyday routes, through the languages of façades and the choreography of markets. They remember because people inhabit them, tell their stories there, and recreate them every day. This book grew from a conviction that design can do more than repair or embellish historic fabric: it can repair and create the social bonds that make heritage places meaningful and resilient.

Urban Acupuncture for Place Attachment and Social Resilience in Heritage Contexts is an inquiry into that conviction. It asks a simple but consequential question: can small, carefully targeted design interventions strengthen the psychological and social foundations that allow communities to survive and thrive under pressures of gentrification? To answer this, the book brings together architectural psychology, environmental psychology and urban design practice into a single, operational framework. The core argument is that place attachment mediates the effect of urban design on social resilience. Urban acupuncture, conceived as small, place-sensitive interventions, becomes effective to the extent that it restores meaning, routine and stewardship in heritage environments.

The research presented here is intentionally comparative and interpretive. Three case studies—spanning a proposed waterfront strategy (LibreLigne, Libreville), an implemented passageway retrofit (Kodak Passageway, Cairo), and an urban-network intervention (Al-Rusafa, Baghdad)—provide contrasting lenses through which the framework is tested.

Readers will find four tightly connected chapters. Respectively, the first and second chapters set the methodological and theoretical groundwork by reviewing the literature on urban acupuncture, place attachment, and social resilience, and explain the mediation logic that structures the analysis. The third chapter presents three case studies in parallel, using a consistent set of parameters that includes coping, adaptive and transformative capacities; multi-functionality, flexibility and predictability, as well as proxy indicators of emotional, cognitive and behavioral bonds. The findings are compiled and synthesized in the fourth chapter, along with a discussion and conclusion, design-oriented recommendations that help assure equitable outcomes in heritage contexts, followed by a book summary.

This book is intended for practitioners and scholars alike: for architects and urban designers who want design strategies with social intent; for heritage professionals seeking ways to keep places alive; for planners and policy makers who must balance continuity and change; and for researchers interested in operationalizing psychological constructs in built-environment inquiry. This book is an argument for a more modest and humane urban design—one that values memory alongside function, and that sees the act of designing as an invitation to belong.

Beirut, Lebanon

Baher I. Farahat
Helga O. Alaeddine

Competing Interests The authors have no competing interests to declare that are relevant to the content of this manuscript.

Contents

List of Figures

List of Tables

Chapter 1
Introduction

Cities are not only assemblies of buildings and streets—they are lived environments in which psychological processes shape how people feel, remember, and act. Architectural psychology and environmental psychology make that link explicit: they examine how design influences emotion, cognition, and behavior, and conversely how human attachment, identity and collective memory demand specific spatial responses. This book foregrounds that reciprocity by treating design as a psychological intervention—what we call an **architecture of belonging**—where material choices, spatial legibility and programmed practices work together to cultivate emotional bonds, meaningful place-memory, and everyday stewardship. Framing urban acupuncture through this lens shows how small, targeted interventions can repair the social fabric of historic neighborhoods by restoring the conditions for belonging rather than merely reshaping physical form.

Historic cities carry the weight of collective memory, identity, and belonging. Over centuries, alleyways and plazas accumulate stories of daily life, cultural exchange, triumph, and loss. Yet today, many of these districts face an invisible wound: gentrification. As high-end developments creep in, rising rents push out long-time residents; traditional businesses close; vernacular streetscapes give way to glass facades. In this process, a community's emotional bond to place—its place attachment—frays, and with it, the social resilience that once helped neighborhoods endure shocks.

The architecture of belonging refers to the intentional design of spaces that foster emotional connection, identity, and inclusivity, forming a key principle in architectural psychology. Rooted in architectural psychology, this concept emphasizes how built environments influence human behavior, attachment, and well-being. Research has shown that the built environment plays a critical role in shaping human experience, influencing how individuals feel connected to place and community (Gifford, 2007). Spaces that reflect cultural identity, historical continuity, collective memory and local narratives contribute to stronger place attachment and promote psychological well-being (Manzo & Devine-Wright, 2013). The sense of belonging that

B. I. Farahat and H. O. Alaeddine, *Urban Acupuncture for Place Attachment and Social Resilience in Heritage Contexts*, Creativity, Heritage and the City 10,
https://doi.org/10.1007/978-981-95-8970-8_1

emerges from such environments supports social cohesion and resilience, particularly in contexts of displacement or urban transformation (Scannell & Gifford, 2010).

When people feel that a space reflects who they are and accommodates their needs, they are more likely to engage with it meaningfully, reinforcing both personal identity and communal bonds. Thus, designing for belonging is fundamental to creating sustainable and resilient urban environments. In this regard, architecture becomes a tool not only for shelter but also for cultivating identity and emotional security within a changing urban fabric.

The architecture of belonging expands on this diagnosis of cultural disintegration by suggesting urban acupuncture as a precise, human-centered alternative. Community-driven initiatives can restore daily social routines and rekindle collective memory by focusing on the most vulnerable nodes in the urban fabric. These micro-projects not only repair physical decay but also rekindle place attachment, reinforcing residents' emotional investment in their neighborhood. In doing so, they lay the groundwork for renewed social resilience, enabling communities to adapt, recover, and thrive even as broader economic and demographic forces press in. The chapters that follow will first unpack the theoretical foundations of gentrification, place attachment, and social resilience, and then present a practical methodology for applying urban acupuncture, before illustrating its impact through detailed case studies and actionable policy insights.

This book proposes a precise and human-centered remedy. Drawing on the metaphor of acupuncture—a practice of small, targeted interventions that restore circulation and balance—the researchers argue that **urban acupuncture** can reconnect people to their evolving heritage landscapes. By intervening at key nodes—an overlooked courtyard, a disused fountain, a crumbling corner façade—designers, residents, and stakeholders can spark renewed emotional investment in place. These micro-interventions do more than repair brick and mortar; they repair the social fabric. This book offers a multidimensional exploration of how urban acupuncture can restore place attachment and foster social resilience in gentrified heritage contexts and neighborhoods. It presents a strong theoretical foundation linking spatial displacement to the erosion of emotional bonds with place, and outlines a clear methodology for implementing targeted, community-driven interventions. Through specific case studies, it demonstrates how small-scale design actions can revive collective memory, enhance community cohesion, and catalyze sustainable stewardship. Ultimately, it argues that the strength of urban acupuncture lies in its capacity to restore belonging—empowering communities to care for, reimagine, and coexist with their heritage in meaningful ways.

Building on this diagnostic framework, the book' chapters are centered in a scientific research ensuring that case studies and policy analyses reflect the latest evolutions in urban and heritage practice. The research employs a robust qualitative methodology—combining interviews, focus groups, and comparative case studies—to capture residents' lived experiences, unpack the "how", and "why" behind each intervention's impact. Central to our inquiry is a four-stage conceptual model (gentrification pressure → urban acupuncture → place attachment →

social resilience), from which we derive testable hypotheses: namely, that targeted micro-interventions bolster place attachment, which in turn mediates gains in social resilience. Moreover, this book offers actionable diagnostic tools, policy recommendations, and design guidelines tailored for planners and heritage managers. Although illustrated through projects in African and Middle Eastern quarters, the micro-intervention strategies presented here are designed for universal applicability, offering a transferable blueprint for revitalizing heritage worldwide.

By the end of this exploration, we will recognize that the true power of urban acupuncture lies not in grand masterplans, but in restoring the simple acts of belonging, of people gathering, remembering, and caring for the places they call home. In doing so, the book offer a fresh lens on how designers, planners, and residents can work together to heal historic urban environments, one precise intervention at a time.

1.1 Research Approach

Heritage districts are increasingly threatened by gentrification—a process whereby rising investment and shifting demographics erode local identity, displace long-standing residents, and weaken the emotional bonds that tie people to place (Lees et al., 2008; Smith, 1979). Concurrently, research in environmental psychology has demonstrated that **Place Attachment**—the affective bond between individuals and specific settings—is critical to community cohesion and the capacity to withstand socio-spatial disruptions (Low & Altman, 1992; Lewicka, 2011). **Social resilience**, understood as the ability of communities to absorb, adapt to, and recover from shocks, likewise depends on these deep ties of memory, identity, and belonging (Norris et al., 2008). On the other hand, **Urban acupuncture**—the practice of deploying small, precise, and low-cost interventions to stimulate positive change in the urban fabric—offers a targeted strategy for heritage regeneration (Lerner, 2014; Waldheim, 2016). By identifying and activating key nodes—such as a neglected courtyard, a disused fountain, or a fading façade—these micro-interventions can rebuild place attachment and trigger broader social and ecological benefits (Lerner, 2014). This book, The Architecture of Belonging: Urban Acupuncture for Place Attachment and Social Resilience in Heritage Contexts investigates how such interventions can restore emotional investment in historic neighborhoods, strengthen community agency, and foster long-term resilience against the homogenizing forces of gentrification.

Resilience is frequently characterized as the capacity to bounce back from adversity and regain strength, health, or success, or to resume its original form after being pulled, stretched, pressured, or bent by external forces (Trogal et al., 2019). In this context, social resilience refers to a community's ability to absorb disruptions, adapt to change, and maintain its core identity and functioning (Koliou et al., 2018; Peacock et al., 2005; Walker et al., 2004).

This research investigates how gentrification, through the displacement of local residents and erosion of cultural identity, undermines these adaptive capacities in historic communities—particularly those with rich intangible heritage. Promoting culture and creativity is therefore critical, not only for economic sustainability but also for reinforcing community bonds and local stewardship (European Commission, 2017).

One essential yet often overlooked variable in this equation is place attachment—the emotional and cognitive bond individuals and groups form with their built environment, which is vital for fostering resilience and encouraging long-term resident engagement (Lewicka, 2011; Manzo & Perkins, 2006). When these bonds are severed, as in many gentrifying historic areas, the loss of collective memory and identity can destabilize the social fabric. Conversely, architectural interventions—such as those practiced in urban acupuncture—can support the renewal of place attachment by enabling community participation, reactivating underutilized spaces, and preserving symbolic landmarks. As such, this research explores how small-scale, culturally sensitive urban interventions can strengthen both place-based identity and social resilience in heritage contexts.

By weaving together empirical analyses of gentrification's social costs (Chapple & Zuk, 2016; Davidson & Lees, 2005), psychological studies that illuminate the emotional and cognitive foundations of place attachment (Lewicka, 2011; Scannell & Gifford, 2010), and vivid case studies of urban acupuncture in action (Lerner, 2014), this research not only validates the real-world impact of neighborhood change but also uncovers the mechanisms by which small interventions restore residents' bonds to place. Anchored in global policy frameworks—from UNESCO's cultural resilience mandates (UNESCO, 2016) to UN-Habitat's sustainable urbanization goals (UN-Habitat, 2020)—and informed by practical resilience indices such as the Baseline Resilience Indicators for Communities (BRIC) model (Cutter, 2016), this book' approach offers both the evidence and the tools needed to measure, design, and implement heritage-sensitive strategies that rebuild community identity, fortify social networks, and ensure that historic cities can withstand and thrive beyond the pressures of gentrification.

1.2 Problem Definition

Gentrification is the process by which certain geographic areas become more exclusive, which has a negative impact on the elderly, families, young people, and those who live in poverty (Thurber & Krings, 2021a, 2021b). It unfolds through economic, cultural, social, psychological, and environmental shifts: as locals leave and new residents arrive, the place attachment—the emotional and cognitive bond between people and their neighborhood—erodes, undermining both individual well-being and collective identity (Lewicka, 2011; Scannell & Gifford, 2010).

The physical features of gentrification are as follows:

- Wealthier people replace local residents
- Rising rents and property values

- Changes in neighborhood character and scale
- Loss of tangible and intangible heritage

This phenomenon is affecting many cities all over the world on various social and cultural levels, frequently taking place on a neighborhood scale (Maciag, 2015a, 2015b). The historic urban environments—where collective character and scale are central—are losing their identity due to the invasion of new infill projects, or demolition of built heritage for the benefit of developers. By breaking existing social networks, diminishing place attachment and escalating social tensions, gentrification can have a negative impact on psychological well-being by threatening both community cohesion and resilience.

Ultimately, as gentrification disrupts the deep emotional bonds residents hold with their environment, the resulting loss of place attachment not only destroys community identity and well-being but also undermines social resilience. Addressing this challenge requires strategies—such as urban acupuncture—that deliberately restore and reinforce those bonds, ensuring that heritage settings can adapt to change without sacrificing the cultural continuity and collective strength that define them.

1.3 Research Aim

This book aims to demonstrate that strengthening place attachment through targeted urban acupuncture interventions can enhance social resilience and enable communities to effectively confront and mitigate the impacts of gentrification in heritage contexts. It seeks to show that small-scale, context-sensitive spatial interventions—when grounded in the psychological dimensions of place—can reinforce emotional bonds, shared meanings, and everyday practices that tie residents to their historic environments. By doing so, these interventions support communities' coping, adaptive, and transformative capacities, helping them resist socio-spatial displacement, cultural erosion, and identity loss commonly associated with gentrification. Ultimately, the research positions urban acupuncture not merely as a design strategy, but as a psychosocial tool capable of sustaining belonging, continuity, and resilience within culturally significant urban fabrics.

1.4 Conceptual Model: The Urban Acupuncture–Place Attachment–Social Resilience Sequence in Gentrifying Heritage Contexts

To move from theory to empirical testing, we propose a conceptual model in which the intensity of urban acupuncture interventions (independent variable) enhances social resilience outcomes—such as community cohesion, adaptive capacity, and heritage retention (dependent variable)—through strengthened place attachment (mediating variable).

Urban acupuncture is targeted; micro-scale design actions revitalize underused public spaces and foster local stewardship (Lerner, 2014). Place attachment, defined as the emotional, cognitive, and behavioral bond to a locale, channels these interventions into lasting community engagement (Lewicka, 2011; Scannell & Gifford, 2010). Social resilience, the community's capacity to withstand and recover from socio-spatial disruptions, is operationalized using indicators of cohesion, adaptability, and cultural continuity (Norris et al., 2008; Walker et al., 2004).

1.4.1 Gentrification Pressure

Gentrification pressure manifests as a convergence of economic, demographic, and cultural stressors that systematically alter historic neighborhoods. As investment streams into a district, property values and rents rise, often displacing lower-income residents and altering the social composition (Maciag, 2015a, 2015b; Thurber & Krings, 2021a, 2021b).

These shifts frequently entail the demolition or insensitive infill of heritage buildings, eroding both tangible and intangible cultural landmarks (Davidson & Lees, 2005). Consequently, established social networks—characterized by relationships of trust and reciprocity that are foundational to both individual and community well-being (Perkins et al., 1996)—weaken, leaving communities more vulnerable to further disruption and diminishing the shared sense of place that underpins collective resilience.

1. Lost Social Networks

Strong place-based interpersonal networks are frequently the source of friendship, social support, and resource sharing for residents of low-income areas. Due to gentrification and rising housing costs, these ties are becoming more strained (Clampet-Lundquist, 2010). Peer networks within and outside of schools are disrupted for children whose families have moved (Anguelovski et al., 2020). Additionally, older people are more sensitive to disruptions in their social networks because they are more likely to rely on neighborhood-based connections for social support (Torres, 2020). It is true that in theory new connections could be made between long-term and recent residents, though research indicates that this rarely occurs and that they have various, frequently conflicting expectations regarding community life (Copic et al., 2020).

2. Social Tensions

Shifting racial, ethnic, and socioeconomic demography in gentrifying neighborhoods are frequently exacerbate preexisting patterns of discrimination, isolation and social exclusion which are all reported to occur at higher rates by people of color, children, and unhoused individuals (Huyser & Meerman, 2014). Newcomers with higher incomes often face racism, including racial profiling and reduced access to services, as well as subtle forms of exclusion in public spaces and community meetings (Desmond, 2017). This compounded marginalization not only heightens social isolation but also undermines trust in local institutions, weakening the social capital essential for collective resilience (Sandercock & Lyssiotis, 2003). Moreover, children in these communities experience disruptions to their social networks and schooling environments, which can lead to adverse developmental and educational outcomes (Ferguson et al., 2013).

3. Cultural Consequences

The collective habits, traditions, arts, institutions, and social practices of a particular place change because of gentrification, which has a negative impact on the local culture.

Gentrification reshapes not only the built environment but also the cultural fabric of a neighborhood, as long-standing habits, traditions, arts, institutions, and social practices often become marginalized or commodified (Lees et al., 2008). Newcomers' consumption patterns, such as trendy cafés and art galleries, often repackage local customs and community-led events for tourists, compromising their authenticity and accessibility (Zukin, 2009). Moreover, established cultural institutions may shift their programming to cater to wealthier audiences, displacing grassroots festivals and informal social gatherings that once anchored community identity (Shaw & Hagemans, 2015). This cultural displacement undermines the intangible heritage that UNESCO defines as "the practices, representations, expressions, knowledge and skills" vital to a community's sense of self and continuity (UNESCO, 2003).

The weakening of daily rituals and creative expressions leads to a decrease in residents' emotional bonds to their place, reducing their collective capacity to adapt to change.

4. Transformed Place Identity

In conjunction with gentrification, neighborhoods frequently undergo rebranding in an effort to appeal to a new, more affluent clientele. Long-term residents frequently lose their ability to shape the character of their neighborhood as municipal elites force changes to the names of streets, parks, and schools, erasing historical connotations connected to particular neighborhoods (Chidester & Gadsby, 2009). Changes in place identity are reflected in the built environment, the commercial environment, and the people who have access to these areas. In gentrifying neighborhoods, disputes over who should have access to public spaces like bike lanes and dog parks frequently arise between long-time residents and more recent arrivals (Lubitow et al., 2016). Changes in a neighborhood's place identity—the shared meanings and symbolic associations attached to that locale—reshapes the boundaries of who is perceived as "in place" versus "out of place." When a neighborhood is defining symbols, demographics, or built form shift, long-time residents and newcomers may each feel "out of place," eroding shared meanings and leading to social exclusion; without deliberate, inclusive interventions to renegotiate place identity, these dynamics undermine community cohesion.

5. Diminished Place Attachment and Sense of Belonging

The foundation of both individual and societal well-being is the sense of belonging. People who experience a stronger sense of place are more inclined to become involved in their communities, particularly when they see a threat to that place, and this relationship exists between place attachment and civic involvement. Locals who are relocated could feel less connected to their previous community, while locals who stay would have less locations to call home. The absence of third spaces—places other than one's residence and place of employment—that promote a sense of community is problematic, especially for young and old people who rely on these meeting places to meet their social requirements. Third spaces include places like parks, coffee shops, and libraries (Anguelovski et al., 2020; Schusler & Krings, 2018; Torres, 2020).

In response to these compounded stressors, urban acupuncture offers a strategic countermeasure: small-scale, precisely targeted design interventions that activate key nodes within the historic fabric to catalyze broader social and spatial renewal (Lerner, 2014). By focusing resources on underused plazas, neglected façades, or community landmarks, these micro-projects can spark ripple effects—reviving everyday practices, reestablishing local gathering points, and strengthening the emotional bonds that underpin place attachment (Waldheim, 2016).

1.4.2 Urban Acupuncture Interventions

Urban acupuncture employs small-scale, precision interventions—"pinpricks" of design—that catalyze broader social, ecological, and cultural renewal within historic neighborhoods and cultural heritage contexts. By identifying underutilized or deteriorating public spaces (e.g., courtyards, pocket parks, façades) and implementing low-cost, community-driven enhancements, practitioners can revitalize daily social interactions and foster collective stewardship (Lerner, 2014). Such interventions not only improve physical amenities but also signal care and investment, strengthening residents' emotional connection to place (Waldheim, 2016). Empirical studies of tactical urbanism projects demonstrate that when locals are engaged in co-design and activation—through pop-up gardens, temporary plazas, or heritage-inspired art installations—these micro-actions often "ripple" outward, prompting further grassroots initiatives and enhancing adaptive capacity (Lydon & García, 2015).

1.4.3 Place Attachment Bonds

Place attachment refers to the emotional, cognitive, and behavioral bonds that individuals and communities form with their environment, encompassing feelings of belonging, identity, and meaningfulness (Lewicka, 2011; Scannell & Gifford,

2010). Urban acupuncture interventions—by restoring familiar landmarks, reactivating communal spaces, and involving residents in co-design—can rekindle these bonds, reinforcing both personal and collective memories of place (Manzo & Perkins, 2006). As residents engage in the stewardship of revitalized courtyards or participate in heritage-inspired art projects, their sense of ownership and emotional investment increases, which, in turn, motivates ongoing care and social interaction (Scannell & Gifford, 2010). Strengthened place attachment thus mediates the effect of targeted design actions on community cohesion and adaptive capacity, serving as the psychological bridge between micro-interventions and broader social resilience.

1.4.4 Social Resilience Outcomes

The concept of resilience derives from the Latin word "*resilio*", meaning "to spring back" or "rebound," and denotes the capacity to adjust to or manage the effects of sudden change (Rose, 2009).

In the context of a community, resilience is defined as the capacity to endure, absorb, cope with and adapt to a variety of environmental and social hazards. "Magis" conceptualizes community resilience as "the existence, development, and engagement of community resources by community members to thrive in an environment characterized by change, uncertainty, unpredictability, and surprise," explicitly linking individual health, social networks, and preparedness to collective outcomes (Magis, 2010).

Community resilience depends on individual health and resiliency because successful communities are made up of individuals who are healthy, linked to their social networks, and well-prepared to tolerate, handle, and recover from disasters.

In the face of these interconnected dynamics—gentrification pressure, targeted urban acupuncture, and the mediating role of place attachment—it becomes clear that the identity of a community is at stake. Indeed, the growth of identity—or, more accurately, the loss of identity—seems to be significantly impacted by gentrification. According to theories of place attachment and place identification, a person's relationship to their immediate surroundings shapes how their identity is created, preserved, and maintained (Cahill, 2007). As gentrification alters the physical environment and disrupts familiar social landscapes, residents experience not only a diminished sense of attachment to place but also an increasing estrangement from the community that once defined them.

1.4.5 *Significance of a Heritage Context*

A heritage context is a geographically defined area where both tangible and intangible assets coalesce around a unifying theme—whether historical, cultural, or architectural—and collectively convey the identity and memory of a community (Australia ICOMOS, 2013; UNESCO, 2011).

Such areas typically include:

- A concentration of buildings, landscapes, or sites whose continuity or development over time reflects shared cultural or historical narratives (ICOMOS, 1999; U.S. Department of the Interior, 2017).
- Places that have witnessed important local, regional, or international events, anchoring collective memory (UNESCO, 2011).
- Associations with individuals—community leaders, artists, architects, or builders—whose work or presence significantly shaped the area's evolution (Australia ICOMOS, 2013).
- Architectural works of recognized quality, displaying masterful design, distinctive construction methods, or the use of indigenous materials (ICOMOS, 1999; U.S. Department of the Interior, 2017).
- Structures or streetscapes exemplifying key architectural styles or construction techniques representative of a particular era (U.S. Department of the Interior, 2017).
- Living traditions, rituals, and social practices that animate the built environment and reinforce community bonds (UNESCO, 2011).
- Character and integrity sufficient to warrant ongoing preservation, restoration, and adaptive reuse (UNESCO World Heritage Centre, 2019).

Together, these dimensions—physical fabric, historical events, personal and artistic associations, material craftsmanship, and enduring cultural practices—define a heritage context as much more than a collection of old buildings: it is a living matrix of place, memory, and identity (Australia ICOMOS, 2013; UNESCO World Heritage Centre, 2019).

1.4.6 *Urban Measure to Enhance Place Attachment and Social Resilience*

Cities are an illustration of complex systems. A number of variables that are constantly changing and evolving affects such a living, dynamic system, with its interconnected parts. Particularly in developing cities, the rate of change is too rapid to fully comprehend the various factors causing it. This explains why cities are such complex topics and why it is challenging to come up with straightforward solutions (Redman, 2014).

Globally, significant progress has been made in addressing emerging hazards, and as a result, the dominant perspective has shifted from a strict emphasis on vulnerability reduction to an improved capacity to withstand disasters. According to this perspective, risk reduction initiatives ought to emphasize resilience thinking and work to develop and strengthen the traits of resilient societies (Rafieian et al., 2011).

Building on the definition of heritage contexts as living matrices of physical fabric, cultural memory, and community practices, we now turn to the specific urban measures that can reinforce those bonds and bolster resilience. Targeted interventions—ranging from revitalizing underused courtyards and façades to co-designing pocket parks and cultural installations—have been shown to restore place attachment by reactivating local gathering points and preserving symbolic landmarks (Lerner, 2014; Waldheim, 2016). In turn, stronger attachment strengthens social resilience, as evidenced by improved community cohesion, adaptive capacity, and heritage stewardship (Norris et al., 2008; Scannell & Gifford, 2010). With this foundation, we can now articulate our hypotheses concerning how urban acupuncture interventions, mediated by place attachment, drive resilient outcomes in gentrifying heritage contexts.

1.5 Research Hypothesis

This research proposes that urban acupuncture can strengthen place attachment and social resilience in heritage contexts where gentrification—and subsequent tourism—has reshaped the neighborhood. By deploying small-scale, community-centered interventions that directly address gentrification's psychosocial effects, this strategy seeks to rebuild residents' emotional bonds to place, reinforce local identity, and mobilize collective capacities for adaptation.

Given the conceptual framework of this research, we consider the order of concepts: Gentrification (the structural pressure) which is the contextual variable—Place Attachment (what is psychologically damaged) which is the mediating variable—Social Resilience (what declines as a consequence)—Urban Acupuncture (the intentional response) which is the targeted design intervention.

Building on the conceptual model (Gentrification → Place Attachment → Social Resilience; Urban Acupuncture as the design intervention), the study tests the following directional hypotheses:

H1—Greater intensity and contextual specificity of urban acupuncture interventions will produce higher levels of place attachment among local residents and users.

H2—Higher place attachment (emotional, cognitive and behavioral) will be associated with improved social resilience outcomes (greater coping, adaptive and transformative capacities).

H3—Place attachment mediates the relationship between urban acupuncture and social resilience: the effect of intervention intensity on resilience operates primarily through its capacity to rebuild place attachment.

1.6 Research Determinants

This research is defined by clear methodological and epistemological boundaries that ensure its relevance, rigor, and coherence. It allows the analysis to engage with the most recent theoretical advances and applied practices related to urban acupuncture, place attachment, and social resilience in gentrifying heritage neighborhoods. The selected case studies and the supporting literature reflect contemporary urban conditions, policy discourses, and design responses shaped by accelerated urban change and tourism-driven pressures.

Methodologically, the research adopts an analytical approach, prioritizing the examination of psychosocial dynamics, particularly place attachment and social resilience, within heritage contexts. Urban acupuncture is examined as a targeted design strategy operating across architectural, urban, and social scales, with its impacts assessed through a defined set of spatial and social parameters. The study employs observational, documentary, and visual proxy indicators, enabling a consistent comparative analysis across diverse geographic and cultural contexts.

Scientific rigor is maintained throughout all stages of the research. The literature review synthesizes peer-reviewed academic sources and established theoretical frameworks, while the case study analyses draw on a combination of primary design documentation and secondary empirical sources. All analytical claims are systematically grounded in scholarly references and cited according to APA standards. Finally, to uphold academic integrity, the manuscript adheres to institutional regulations, ensuring originality, transparency, and ethical scholarly practice.

1.7 Research Methodology

This research employs a qualitative methodology to explore how urban acupuncture interventions foster place attachment and social resilience in gentrifying heritage neighborhoods. Grounded in social and behavioral science traditions, and integrating perspectives from urban planning and architecture, the research prioritizes depth of understanding over breadth: it seeks both the observable changes in space and use and the psychosocial mechanisms linking design to social outcomes. Place attachment is treated as the pivotal mediating construct connecting targeted, micro-scale interventions to community resilience.

Analytically, the research applies inductive coding to identify emergent patterns, comparative analysis to contrast different heritage contexts, and thematic interpretation to connect intervention intensity with levels of place attachment and resilience.

The primary empirical material comprises three comparative case studies selected for contrast in scale and maturity (conceptual design → implemented micro-intervention → urban-network intervention). Data sources include: project documentation and design drawings; policy and planning documents; photographic records and media reports; observational field notes (where available); and stakeholder interview transcripts and focus-group summaries when access permits. Where direct social surveying is limited, rigorously defined proxy indicators (programming records, recurrent use patterns, photographic transects, and local media accounts) are used to infer psychosocial phenomena.

The methods of data collection includes: Documentary and archival analysis such as design proposals, municipal documents and project reports—Observation and photographic transects such as documented time-series photos and site descriptions to capture behavioral patterns and physical change—Content analysis of digital storytelling tools, apps, and interpretive materials associated with interventions.

The analytical strategy involves three main components: thematic interpretation, which groups data into resilience dimensions (coping, adaptive, transformative) and place attachment constructs (emotional, cognitive, behavioral); comparative case analysis that juxtaposes cases based on urban-acupuncture variables, place-attachment proxies, and social resilience outcomes to identify patterns and distinctions; and mediation argumentation that uses triangulated qualitative data and quantitative proxies to evaluate the relationship between intervention intensity, place attachment, and social resilience.

Operational evaluation of social resilience involves defining success through coping, adaptive, and transformative capacities; documenting the design and implementation of urban-acupuncture parameters such as multi-functionality, flexibility, and predictability; and interpreting mechanisms by assessing how place attachment proxies (emotional, cognitive, behavioral) explain the relationship between interventions and resilience outcomes.

It is pertinent to mention that the lack of universal survey data poses challenges in accurately assessing individual experiences, suggesting the need for post-occupancy monitoring. Case heterogeneity limits generalizations, promoting a framework that identifies specific conditions for transfer rather than universal solutions.

Lastly, this cross-disciplinary approach ensures that the study's insights into urban acupuncture's role in cultivating collective adaptability are both empirically grounded and theoretically coherent.

Consequently, the following statement illustrates the evaluation rationale of this book methodology:

1. Define "what success means" → parameters for social resilience
2. Examine "what was done" → criteria of urban acupuncture
3. Interpret "why it worked" → proxy indicators of place attachment

References

Anguelovski, I., Brand, A. L., Connolly, J. J., Corbera, E., Kotsila, P., Steil, J., Lamarca, M. G., Triguero-Mas, M., Cole, H. V. S., Baró, F., Langemeyer, J., del Pulgar, C. P., Shokry, G., Sekulova, F., & Argüelles Ramos, L. (2020). Expanding the boundaries of justice in urban

greening scholarship: Toward an emancipatory, antisubordination, intersectional, and relational approach. *Annals of the American Association of Geographers, 110*(6), 1743–1769.
Australia ICOMOS. (2013). *The Burra Charter: The Australia ICOMOS Charter for places of cultural significance*. Australia ICOMOS.
Cahill, C. (2007). The personal is political: Developing new subjectivities through participatory action research. *Gender, Place and Culture, 14*(3), 267–292.
Chapple, K., & Zuk, M. (2016). Forewarned: The use of neighborhood early warning systems for gentrification and displacement. *Cityscape, 18*(3), 109–130.
Chidester, R. C., & Gadsby, D. A. (2009). One neighborhood, two communities: The public archaeology of class in a gentrifying urban neighborhood. *International Labor and Working-Class History, 76*(1), 127–146.
Clampet-Lundquist, S. (2010). "Everyone had your back": Social ties, perceived safety, and public housing relocation. *City & Community, 9*(1), 87–108.
Copic, C., Schusler, T. M., & Krings, A. (2020). Environmental gentrification in Chicago: Perceptions, dilemmas, and paths forward. *Sustainability, 12*(16), 6536. https://doi.org/10.3390/su12166536
Cutter, S. L. (2016). The landscape of disaster resilience indicators in the USA. *Natural Hazards, 80*(2), 741–758.
Davidson, M., & Lees, L. (2005). New-build 'gentrification' and London's riverside renaissance. *Environment and Planning A, 37*(7), 1165–1190. https://doi.org/10.1068/a37209
Desmond, M. (2017). *Evicted: Poverty and profit in the American city*. Crown.
European Commission. (2017). *Culture and creative sectors boosting structural change in European cities*. Publications Office of the European Union. Retrieved from https://ec.europa.eu/
Ferguson, K. T., Cassells, R. C., MacAllister, J. W., & Evans, G. W. (2013). The physical environment and child development: An international review. *International Journal of Psychology, 48*(4), 437–468.
Gifford, R. (2007). *Environmental Psychology: Principles and Practice* (4th ed.). Optimal Books.
Huyser, M., & Meerman, J. R. (2014). Resident perceptions of redevelopment and gentrification in the heartside neighborhood: Lessons for social work profession. *J. Soc. & Soc. Welfare, 41*, 3.
International Council on Monuments and Sites [ICOMOS]. (1999). *International Charter for the Conservation and Restoration of Monuments and Sites (The Venice Charter)*. ICOMOS. https://www.icomos.org/charters/venice_e.pdf
Koliou, M., van de Lindt, J. W., McAllister, T. P., Ellingwood, B. R., Dillard, M., & Cutler, H. (2018). State of the research in community resilience: Progress and challenges. *Sustainable and Resilient Infrastructure, 5*(3), 131–151. https://doi.org/10.1080/23789689.2017.1418547
Lees, L., Slater, T., & Wyly, E. (2008). *Gentrification*. Routledge. https://doi.org/10.4324/9780203940877
Lerner, J. (2014). *Urban acupuncture: Celebrating pinpricks of change that enrich city life*. Lars Müller Publishers.
Lewicka, M. (2011). Place attachment: How far have we come in the last 40 years? *Journal of Environmental Psychology, 31*(3), 207–230. https://doi.org/10.1016/j.jenvp.2010.10.001
Low, S. M., & Altman, I. (1992). Place attachment: A conceptual inquiry. In I. Altman & S. M. Low (Eds.), *Place attachment* (pp. 1–12). Springer, US.
Lubitow, A., Zinschlag, B., & Rochester, N. (2016). Plans for pavement or for people? The politics of bike lanes on the 'Paseo Boricua' in Chicago, Illinois. *Urban Studies, 53*(12), 2637–2653.
Lydon, M., & Garcia, A. (2015). A tactical urbanism how-to. *Tactical urbanism: Short-term action for long-term change* (pp. 171–208). Island Press/Center for Resource Economics.
Maciag, M. (2015a, February 24). Gentrification in America report. *Governing*. https://www.governing.com/archive/gentrification-in-cities-governing-report.html
Maciag, M. (2015b, September 28). Millennials remade cities. But will they keep living in them? *Governing*. Retrieved from https://www.governing.com/archive/gentrification-in-cities-governing-report.html

Magis, K. (2010). Community resilience: An indicator of social sustainability. *Society and Natural Resources, 23*(5), 401–416.

Manzo, L., & Devine-Wright, P. (2013). *Place attachment: Advances in theory, methods and applications*. Routledge.

Manzo, L. C., & Perkins, D. D. (2006). Finding common ground: The importance of place attachment to community participation and planning. *Journal of Planning Literature, 20*(4), 335–350. https://doi.org/10.1177/0885412205286160

Norris, F. H., Stevens, S. P., Pfefferbaum, B., Wyche, K. F., & Pfefferbaum, R. L. (2008). Community resilience as a metaphor, theory, set of capacities, and strategy for disaster readiness. *American Journal of Community Psychology, 41*, 127–150. https://doi.org/10.1007/s10464-007-9156-6

Peacock, W. G., Brody, S. D., & Highfield, W. E. (2005). Hurricane risk perceptions among Florida's single-family homeowners. *Landscape and Urban Planning, 73*(2–3), 120–135. https://doi.org/10.1016/j.landurbplan.2010.05.001

Perkins, D. D., Brown, B. B., & Taylor, R. B. (1996). The ecology of empowerment: Predicting participation in community organizations. *Journal of Social Issues, 52*(1), 85–110.

Rafieian, M., Rezaei, M. R., Asgary, A., Parhizkar, A., & Shayan, S. (2011). Explaining the concept of resilience and its indexation in community-based disaster management. *J Spat Plan, 15*(4), 20–41.

Redman, C. L. (2014). Should sustainability and resilience be combined or remain distinct pursuits? *Ecology and Society, 19*(2), 37. https://doi.org/10.5751/ES-06390-1902377

Rose, A. (2006). Economic resilience to disasters: Toward a consistent and comprehensive formulation. *Disaster resilience: An integrated approach*, 275–303.

Sandercock, L., & Lyssiotis, P. (Eds.). (2003). *Cosmopolis II: Mongrel cities of the 21st century*. A&C Black.

Scannell, L., & Gifford, R. (2010). Defining place attachment: A tripartite organizing framework. *Journal of Environmental Psychology, 30*(1), 1–10. https://doi.org/10.1016/j.jenvp.2009.09.006

Schusler, T. & Krings, A. (2018). Addressing environmental gentrification: Improving environmental health for children and youth without displacement. Chicago: Loyola University School of Law Center for the Human Rights of Children.

Shaw, K. S., & Hagemans, I. W. (2015). 'Gentrification without displacement and the consequent loss of place: The effects of class transition on low-income residents of secure housing in gentrifying areas. *International Journal of Urban and Regional Research, 39*(2), 323–341.

Smith, N. (1979). Toward a theory of gentrification a back to the city movement by capital, not people. *Journal of the American Planning Association, 45*(4), 538–548.

Thurber, A., & Krings, A. (2021a). *Gentrification*. Oxford University Press. https://doi.org/10.1093/acrefore/9780199975839.013.1413

Thurber, A., & Krings, A. (2021b). The dynamics of urban exclusivity: Gentrification's impacts on vulnerable populations. *Urban Geography, 42*(5), 700–720.

Torres, S. (2020). "For a younger crowd": Place, belonging, and exclusion among older adults facing neighborhood change. *Qualitative Sociology, 43*(1), 1–20.

Trogal, K., Bauman, I., Lawrence, R., & Petrescu, D. (2019). *Architecture and Resilience: Interdisciplinary dialogues*. Routledge.

UNESCO. (2003). *Convention for the safeguarding of the intangible cultural heritage*. UNESCO. https://ich.unesco.org/en/convention

UNESCO. (2011). *Recommendation on the historic urban landscape*. United Nations Educational, Scientific and Cultural Organization. https://whc.unesco.org/en/hul/

UNESCO. (2016). *Culture: Urban future. Global report on culture for sustainable urban development*. UNESCO Publishing. https://unesdoc.unesco.org/

UNESCO World Heritage Centre. (2019). *Operational guidelines for the implementation of the World Heritage Convention*. UNESCO.

UN-Habitat. (2020). *World Cities Report 2020: The value of sustainable urbanization*. United Nations Human Settlements Programme.https://unhabitat.org/world-cities-report-2020-the-value-of-sustainable-urbanization

U.S. Department of the Interior, National Park Service. (2017). *How to Apply the National Register Criteria for Evaluation* (National Register Bulletin No. 15). https://www.nps.gov/subjects/nationalregister/upload/NRB-15_web508.pdf

Waldheim, C. (2016). *Landscape as urbanism: A general theory*. Princeton University Press.

Walker, B., Holling, C. S., Carpenter, S. R., & Kinzig, A. (2004). Resilience, adaptability and transformability in social–ecological systems. *Ecology and Society, 9*(2), 5.

Zukin, S. (2009). *Naked city: The death and life of authentic urban places*. Oxford University Press.

Chapter 2
Literature Review

In the face of escalating global urbanization, heritage neighborhoods are increasingly vulnerable to processes of gentrification, often resulting in the displacement of long-term residents, erosion of cultural identity, and degradation of socio-spatial continuity. These transformations challenge not only the physical fabric of cities but also the psychological well-being and resilience of communities.

This chapter provides a comprehensive review of the theoretical and empirical literature relevant to the research. It establishes the theoretical foundation upon which the research is constructed by critically examining key bodies of literature at the intersection of architecture, urban design, and environmental psychology. It situates the study within contemporary debates on gentrification, place attachment, social resilience, and urban acupuncture, with particular attention to heritage contexts where socio-spatial change is both accelerated and contested.

The literature review is structured to move from structural pressures to human experience, and finally to design responses. It begins by examining gentrification as a dominant force reshaping historic urban environments, often resulting in displacement, commodification of heritage, and erosion of everyday social life. While existing scholarship has extensively documented the economic and spatial dimensions of gentrification, fewer studies engage deeply with its psychological and social consequences, particularly in relation to belonging, identity, and collective resilience in heritage neighborhoods.

The chapter then turns to place attachment, drawing on environmental and social psychology to understand how emotional bonds, cognitive meanings, and behavioral practices connect individuals and communities to place. Place attachment is reviewed not only as an outcome of long-term habitation, but also as a dynamic process, that can be weakened or strengthened through spatial transformation. This perspective positions place attachment as a critical mediating concept linking urban change to social outcomes.

Building on this psychological grounding, the literature on social resilience is explored as a framework for understanding how communities cope with, adapt to,

B. I. Farahat and H. O. Alaeddine, *Urban Acupuncture for Place Attachment and Social Resilience in Heritage Contexts*, Creativity, Heritage and the City 10,
https://doi.org/10.1007/978-981-95-8970-8_2

and transform in the face of disruption. In heritage contexts, resilience is increasingly understood as a socially embedded capacity, reliant on collective memory, social networks, and continuity of everyday practices rather than physical robustness alone. Architectural and urban studies that foreground resilience through social processes, rather than purely technical solutions, are given particular emphasis.

Finally, the review introduces urban acupuncture as a design-led strategy capable of operating at the scale where psychological and social processes are most tangible. Drawing on theoretical and applied literature, urban acupuncture is examined as a form of targeted, small-scale intervention that can recalibrate everyday spatial experience, reinforce local identity, and support socially resilient outcomes without resorting to large-scale redevelopment.

Grounded in interdisciplinary frameworks from urban planning, architectural psychology, and social studies, this review explores how urban acupuncture can serve as a tool to mitigate the psychosocial impacts of urban change. It critically examines how place attachment functions as a mediating variable, fostering emotional and cognitive bonds between individuals and their environments, and how such bonds can be intentionally reinforced to support social resilience. The literature also reflects on the architectural and cultural significance of heritage contexts, which serve not only as physical environments but also as storied landscapes embedded with memory, identity, and belonging.

By synthesizing these four strands of literature, this chapter identifies the need for an integrated framework that explicitly links gentrification pressures, place attachment, social resilience, and urban acupuncture within heritage contexts. Addressing this gap provides the conceptual rationale for the analytical framework and case study methodology developed in the following chapter.

2.1 Gentrification in Heritage Contexts

Heritage environments are not merely collections of old buildings; they are repositories of cultural identity, collective memory, and architectural legacy. However, these areas have increasingly become hotspots for gentrification, a process wherein rising investment and redevelopment efforts attract wealthier residents, thereby displacing long-standing, often vulnerable populations. In these settings, gentrification manifests not only through physical transformations—such as the renovation of heritage buildings or the insertion of modern infill projects—but also through the erosion of intangible cultural heritage and social networks that define the neighborhood's identity (Smith, 2002; Zukin, 2010).

The transformation is often driven by market-oriented urban policies that prioritize economic growth over cultural continuity (Lees et al., 2008). As a result, long-established communities are replaced or fragmented, and vernacular architectural elements are either lost or commodified. This process undermines both the tangible and intangible dimensions of place, contributing to the alienation of residents and the weakening of social cohesion (Davidson, 2009). Particularly in heritage contexts, the impact is twofold: while restoration and investment may preserve or enhance the physical fabric, the sociocultural ecosystem often suffers irreversible damage.

Moreover, gentrification in historic areas raises complex tensions between preservation and progress. While some redevelopment projects aim to revitalize underutilized or decaying spaces, they may inadvertently exclude original populations and dilute local culture, effectively turning historic districts into sanitized consumer landscapes (Shaw, 2008). These outcomes highlight the urgent need for urban strategies that balance development with the protection of social and cultural continuity—an imperative this research addresses through the lens of place attachment, social resilience, and urban acupuncture.

Figure 2.1 shows the visual and socio-spatial contrasts created by gentrification in a historic urban fabric are demonstrated by a 1920s neoclassical house in Mexico City's Colonia Roma Norte neighborhood next to a contemporary loft tower that is undergoing gentrification and repopulation as of 2000. This contrast shows how historic structures endure in once-stately communities despite the demands of modern renovation. This juxtaposition exemplifies how heritage buildings persist amid contemporary redevelopment pressures in once-stately neighborhoods.

Fig. 2.1 Photo from Mexico City's Colonia Roma: a 1920s neoclassical house alongside a new loft tower, illustrating how a historic district can be altered by gentrification. *Source*: Peláez (2016)

2.1.1 Definitions and Causes of Gentrification

Gentrification is a multifaceted urban process in which reinvestment and market-driven upgrading of previously lower-income neighborhoods attract higher-income residents and capital, producing physical transformation (renovation, infill, redevelopment), socio-economic displacement (rising rents and property values), and cultural change (shifts in local commerce, practices, and identity). Crucially, gentrification involves both material change to the built environment and symbolic or psychosocial displacement—an erosion of residents' place attachment, social networks, and sense of belonging—which together undermine community cohesion and resilience (Davidson, 2009; Lees et al., 2008; Lewicka, 2011; Smith, 1979; Zukin, 2010). The process disproportionately affects vulnerable groups, including low-income households, people of color, and older residents, and raises tensions between heritage preservation and market-led urban renewal (Maciag, 2015a, 2015b; Thurber & Krings, 2021).

Therefore, gentrification is a multifaceted urban process whereby reinvestment and upgrading of deteriorated urban spaces attract higher-income populations, leading to rising property values, cultural shifts, and the displacement of long-term, often marginalized residents. This transformation frequently disrupts place attachment, weakens social networks, and erodes collective identity, especially in historically significant neighborhoods.

Gentrification is driven by a combination of market, policy, socio-cultural and infrastructural forces that together reshape neighborhoods' physical fabric and social composition:

Capital investment and market dynamics: Inflows of private capital and market-led redevelopment (renovation, infill, upscale amenities) raise land and property values and create pressure to replace lower-income residents. (Lees et al., 2008; Smith, 1979).

Housing supply and affordability constraints: Tight housing markets and limited affordable supply amplify rent increases when higher-income demand rises, accelerating displacement (Glaeser, 2012).

Public policy and planning decisions: Zoning changes, tax incentives, redevelopment programs, and pro-growth planning can catalyze or enable gentrification by making investment more attractive (Hackworth & Smith, 2001; Lees et al., 2008).

Infrastructure and amenity upgrades: Transport projects, streetscape improvements, and new cultural or commercial amenities increase an area's attractiveness to wealthier residents and investors (Davidson, 2009).

Lifestyle and consumption preferences: Shifts in preferences toward urban living, cultural amenities, and symbolic authenticity attract higher-income, often younger residents (Florida, 2002; Zukin, 2010).

Tourism and heritage commodification: The commercialization of local culture and heritage for tourists and new consumers can transform local economies and displace traditional practices and residents (Zukin, 2010).

Financialization of housing and global capital flows: Treating housing as a financial asset (investment vehicles, REITs, cross-border capital) increases speculative pressure on urban property markets (Aalbers, 2016).
Demographic and social change: Population turnover, in-migration of more affluent or different sociocultural groups, and changing household structures alter neighborhood social composition and power relations (Lees et al., 2008; Thurber & Krings, 2021).

2.1.2 Consequences of Gentrification

Gentrification produces a constellation of socio-spatial effects—altering the built fabric, local economies, social networks, and everyday meanings of place—that collectively undermine community cohesion and place attachment (Lees et al., 2008; Zukin, 2010). Identifying these key consequences and their measurable indicators is therefore essential for diagnosing vulnerability in heritage contexts and for designing targeted urban-acupuncture interventions that can restore attachment and bolster social resilience.

Gentrification commonly produces residential or physical displacement; whereby rising rents, property values, and redevelopment pressure force long-term, lower-income residents to relocate—fracturing household stability and social networks that sustain everyday life (Lees et al., 2008; Smith, 1979). Displacement is not only an economic outcome but also a social one: eviction, involuntary moving, and tenant churn reduce neighborhood continuity, weaken informal supports, and increase vulnerability among marginalized groups (Desmond, 2017; Thurber & Krings, 2021). Concurrently, cultural or symbolic displacement occurs when the meanings and practices that give a place identity such as local shops, rituals, languages, festivals, and everyday uses of space—are marginalized, commodified, or replaced by consumer-oriented uses serving newcomers and tourists (Zukin, 2010). Even residents who remain may experience a loss of belonging and place attachment as the familiar cultural landscape is altered, eroding collective memory and intangible heritage (Lewicka, 2011; UNESCO, 2003). This form of displacement can be as damaging to social resilience as physical eviction because it undermines the emotional and ritual practices that maintain community cohesion.

Historic and heritage neighborhoods concentrate both tangible (built fabric, streetscapes) and intangible (rituals, crafts, collective memory) values, making them distinct sites of cultural meaning and identity. When gentrification arrives, these areas face a double-edged transformation: market forces and tourism can preserve façades while simultaneously commodifying or displacing living traditions, accelerating both physical eviction and symbolic loss of belonging (Lees et al., 2008; Zukin, 2010). Preservation frameworks and heritage policy further complicate these dynamics—offering tools for protection but also, if misapplied, enabling "museumification" that raises property values and marginalizes long-term residents—thus intensifying threats to place attachment and community identity (Davidson, 2009; UNESCO, 2011).

In addition to the cultural significance conveyed by the loss of intangible heritage, it is essential to consider the effects of gentrification on community identity and place attachment. Strong attachment to place is threatened. Heritage communities often have deep place-based identities. Intangible heritage (as UNESCO notes) "provides communities with a sense of identity and continuity" (https://ich.unesco.org).

Gentrification severs that continuity: when families and neighbors who built the community are expelled, their shared sense of history and belonging erodes. Gentrification anywhere breaks attachments. Long-term residents in any neighborhood (heritage or not) develop loyalties to place. When gentrification accelerates, these ties are broken. What differs is that in heritage areas, the narrative of place (history, architecture, and symbol) becomes entwined with identity.

Therefore, gentrification weakens neighborhood viability and social cohesion through physical displacement, upscale infill, commodification of local economies, social stratification, eroded informal support networks, and securitization of public spaces. This leads to a focus on heritage-specific dynamics, creating challenges and opportunities for interventions like urban acupuncture, which can help address the loss of tangible and intangible practices. Intangible heritage constitutes a community's identity by anchoring social meanings and everyday life (Lewicka, 2011; UNESCO, 2003). In heritage contexts, this deep cultural value also makes neighborhoods especially vulnerable to cultural displacement, since market-driven preservation and tourism can commodify or erase living traditions even when physical fabric is conserved (Lees et al., 2008; Zukin, 2010). Intangible heritage, as defined by UNESCO, is the living practices and expressions that give communities identity and continuity. However, gentrification can lead to the loss or commodification of these traditions, as consumption-driven redevelopment often overlooks original folklore and lifestyle, forcing original residents to relocate.

2.1.3 Heritage Specific Dynamics

Historic and heritage contexts experience gentrification differently because their value is encoded not only in market terms but also in layered meanings—tangible fabric (architecture, streetscapes) and intangible practices (rituals, crafts, memories) that jointly constitute place identity (Lees et al., 2008; UNESCO, 2003). Consequently, gentrification in these areas tends to produce cultural displacement as well as physical displacement: while investment may conserve or beautify façades, it often commodifies living traditions and everyday practices, transforming them into consumable heritage for visitors and new consumers and thereby eroding the community's lived meanings (Lewicka, 2011; Zukin, 2009).

Preservation and heritage policy create a distinct policy environment that can either mitigate or accelerate these effects. Inclusive heritage management—approaches that integrate social equity; affordable housing, and local participation—can protect residents and sustain intangible practices (UNESCO, 2011). However, heritage branding and preservation regulations may also increase property values, decrease the supply of housing, or turn districts into "museumified" areas that put visitors and investors ahead of locals, which would worsen displacement (Davidson, 2009; Waights, 2019). Thus, heritage policy is double-edged: it offers mechanisms for protection but can unintentionally intensify market pressures if not paired with social safeguards.

At the psychosocial level, heritage contexts show intensified impacts on place attachment and community identity. Residents in historic districts typically possess strong place-based attachments rooted in collective memory and routines; gentrification that reconfigures the social and material landscape disrupts these bonds and produces feelings of estrangement and loss of belonging (Lewicka, 2011; Perkins et al., 1996).

Compared with non-heritage neighborhoods, where identity may be more contingent or recent, heritage areas suffer a compounded rupture—loss of occupants combined with the loss of the historical narrative that underpins communal meaning—making recovery of social resilience more complex.

For intervention and research, these dynamics imply that design and policy responses must simultaneously address material conservation and living culture. Strategies such as targeted, community-led micro-interventions (urban acupuncture), combined with heritage-sensitive housing and participatory governance, hold promise for restoring place attachment and buffering social resilience, but they must be explicitly calibrated to preserve intangible heritage and local livelihoods to avoid mere aesthetic preservation that excludes residents (Lerner, 2014; UNESCO, 2011).

2.2 Place Attachment

Place attachment describes the emotional and psychological bonds people form with specific places. It is a key element of community resilience because when residents feel attached and invested in a neighborhood, they are more likely to work together to preserve or rebuild it. Researchers characterize place attachment along three dimensions: an affective (emotional) bond with a location, cognitive connections (memories and meanings associated with the place), and behavioral components (such as activities that reinforce one's ties to the place).

Place attachment endows places with personal and collective significance through individual memories and shared community connections. This concept is crucial for urban designers, as it represents a vital interaction between people and their environments.

2.2.1 *Foundations and Theories from Architectural Psychology*

Place attachment denotes the emotional, cognitive and behavioral bonds people form with particular places; it is a multidimensional concept that connects individual identity and group memory to the material and social qualities of environments (Lewicka, 2011; Scannell & Gifford, 2010).

Positive sentiments that a person has for a certain place originate from the physical surroundings, which house life events. We refer to this emotion as "place attachment". People develop an emotional bond with their typical places, where they feel secure and comfortable and have a strong want to stay close (Altman & Low, 1992). A complex phenomenon includes positive emotions, connectedness, knowledge, behaviors, beliefs, and actions. It promotes experimentation, self-assurance, and freedom in expressing emotions, behaviors, and exploration. It is connected to the availability of privacy, control, safety, comfort, and enjoyment in locations, which fosters emotional ties and a sense of pride and place specific belonging (Hammoodi & Al-Hinkawi, 2023).

Contemporary research organizes place attachment into three interrelated components:

- Emotional (affective) bonds: feelings of love, comfort, security and belonging toward a place. These affective ties reduce stress and motivate stewardship. (Lewicka, 2011; Scannell & Gifford, 2010).
- Cognitive bonds (place identity & meaning): the ways place becomes part of self-concept and collective narratives: historical memory, symbolism, and mental maps that people use to make sense of their lives. This aligns with place-identity theory (Proshansky et al., 1983) and cognitive mapping.
- Behavioral bonds (place dependence and action): the routines, practices and behaviors (use of space, participation, care) that demonstrate commitment to and reliance on a place (Williams & Vaske, 2003; Manzo & Perkins, 2006).

Empirical studies show these dimensions are distinct but mutually reinforcing: emotional attachment motivates behavioral stewardship, while cognitive narratives (stories, histories) sustain the affective sense of home. Measurement approaches therefore combine psychometric scales (validated place-attachment surveys) with behavioral observation and qualitative methods to capture the full tripartite construct (Williams & Vaske, 2003; Scannell & Gifford, 2010).

These three strands interact: emotional attachment motivates behavior; cognitive meanings explain why a place matters; behavior reinforces emotion and identity through ritual and stewardship.

There are numerous models that have been put forth to describe place attachment. However, most scholars agree that the tripartite conceptual framework of attachment to place, which was established by "Scannell and Gifford" (2010), is one of the most prevalent models that meaningfully covers the definitions found in earlier literature (Fig. 2.2).

Fig. 2.2 Place attachment components based on Scannell and Gifford (2010). *Source*: https://mindfolio.com/place-attachment-insight-system-to-create-lasting-communities

According to this paradigm, place attachment is a multifaceted phrase that encompasses the place, the psychological process, and the individual dimensions. It suggests that an individual's attachment to a particular place is a function of the activities and social interactions that take place there and give it significance, making the space a place where the individual feels at home (Bonaiuto et al., 2003; Lewicka, 2011).

Figure 2.2 represents the components of Place Attachment that refers to the emotional bond between people and specific places. Scannell and Gifford (2010) proposed a tripartite framework for understanding place attachment, which can be linked to the components shown in the image. Accordingly, place attachment is multidimensional, involving economic, social, environmental, and symbolic factors that together create a strong emotional connection between people and their environment.

In urban studies and architecture, place attachment encourages individuals to interact with urban places and open spaces. Researchers in a variety of fields, including sociology, environmental psychology, and human geography, have come to understand that place attachment is a crucial social construct for understanding the phenomena that tie people to a particular place (Kyle et al., 2004, 2005).

Place attachment has been defined in several of ways due to its extensive research. A three-dimensional, person-process-place organizing framework is created by reviewing and synthesizing the concept's various interpretations. The person dimension of place attachment refers to its individually or collectively determined meanings. The psychological dimension includes the affective, cognitive, and behavioral components of attachment. The place dimension emphasizes the place characteristics of attachment, including spatial level, specificity, and the prominence of social or physical elements (Scannell & Guilford, 2010).

According to the person–process–place (PPP) framework, place attachment is a bond between an individual or group and a place that can vary in terms of spatial level, degree of specificity, and social or physical features of the place, and is manifested through affective, cognitive, and behavioral psychological processes. However, the question of why people develop such enduring psychological bonds with place remains. Most likely, place attachment bonds exist because they serve several needs and functions.

The most common functions of place attachment include survival and security, goal support (facilitating specific activities), and providing a sense of continuity for personal or collective identity through memory and stability.

- Survival and Security: Places provide a sense of security and safety, which is a foundational aspect of attachment.
- Goal Support: People become attached to places that support their goals and preferred activities, also known as place dependence.
- Continuity: Place attachment provides continuity of memories from the past to the present, supporting personal or social identity.
- Well-being: These bonds improve quality of life and subjective well-being by fostering positive emotions, environmental mastery, and social cohesion (Scannell & Guilford, 2010).

Figure 2.3 visualizes the tripartite model of Place Attachment (person, place, process), which explains the psychosocial construct linking affective ties, meaning making and place-based practices.

The following are the characteristics of place attachment:

- Multiscalar: operates at individual, household, neighborhood and city scales (Lewicka, 2011).
- Dynamic and relational: develops through experience, memory and social interaction—changeable with life events or urban transformation. Culturally embedded: shaped by both tangible heritage (buildings) and intangible practices (festivals, crafts).
- Behaviorally expressed: shown via place-protective actions (stewardship, participation) and everyday routines.

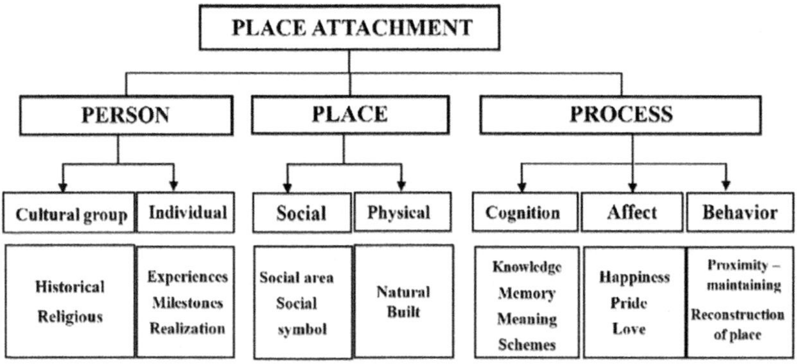

Fig. 2.3 Dimensions of place attachment. *Source*: Al-Mendilawi and Al-Saaidy (2024)

- Predictive of pro-social outcomes: stronger attachment correlates with civic engagement, lower out-migration intentions, and participation in local governance (Manzo & Perkins, 2006).

Place attachment becomes actionable in design and analysis by the conceptual tools provided by architectural psychology:

- Place-identity theory (Proshansky et al., 1983): links environmental settings to self-concept and community identity; it useful for reading heritage as identity scaffolding.
- Affordance and environmental perception (Gibsonian ideas adapted in environmental psychology): show how physical elements invite particular uses and social interactions that sustain attachment (Gifford, 2007).
- Cognitive mapping and legibility: explains how coherent urban form supports wayfinding, memory, and identity.
- Social capital & community participation frameworks (Putnam; Perkins et al., 2002): emphasize trust and reciprocity as behavioral enablers of collective place care (Perkins et al., 1996).
- Emotion-informed design (architectural psychology research): grounds how materiality, scale, and sensory qualities produce wellbeing, belonging and restorative effects.

These theories let architects and planners translate psychological constructs into design principles: visible continuity, affordances for everyday interaction, legible landmarks, and participatory processes.

2.2.2 Role of Place Attachment in the Community Resilience

The field of environmental psychology holds an important position in the research of place attachment. Scientists have characterized place attachment in terms of dimensions, concepts, and components using a variety of themes. The concept of place attachment is diverse and cross-disciplinary. It alludes to the favorable interaction between the person and the environment as shown in the person's emotional, perceptual, and cognitive traits. Positive interactions and activities between people and the location, on the one hand, and the people among themselves, on the other, give the place its significance and meaning. In addition to community involvement and interpersonal connection, this relationship is characterized by preserving proximity to the attachment site, actions, and emotions that are supportive of the urban environment and society (Al-Mendilawi & Al-Saaidy, 2024).

Place attachment is a psychological mechanism that underpins several resilience processes: it motivates collective action, sustains social networks, and fosters investment in local adaptive strategies (Magis, 2010; Norris et al., 2008). Empirically, communities with higher place attachment show greater willingness to rebuild after shocks, participate in recovery, and defend local heritage, thus improving cohesion, adaptability, and recovery (Norris et al., 2008; Walker et al., 2004).

In heritage contexts, attachment to place identity can be a major asset for bottom-up conservation and social resilience. Place attachment promotes social cohesion, civic participation, and informal mutual support, pillars of community well-being and adaptive capacity (Magis, 2010; Norris et al., 2008). Residents with strong attachments are more likely to invest labor and resources in local upkeep, participate in collective problem solving, and engage in place-protecting behaviors (Manzo & Perkins, 2006). In heritage contexts, attachment additionally anchors the transmission of intangible practices (festivals, crafts, oral histories), serving as a psychological resource that communities draw on when confronted with stressors (Folke, 2006). Thus, strengthening place attachment is not merely sentimental: it is an operational lever for enhancing cohesion, supporting learning, and accelerating recovery after disturbance.

2.2.3 Disruption Through Socio-Spatial Change

As human civilizations have grown and people's lifestyles have evolved in response to environmental and technological improvements, place attachment has become more significant in urban and architectural design. This has led scholars to investigate how people's physical and social urban settings affect architects and planners. In order to improve and develop the quality of the urban environment, urban designers and architects began investigating the gaps in the domains of urban design and how these concepts and feelings towards urban locations effect people's perceptions and attitudes towards these places.

Following the incident and devastation of the "Pruitt Igoe" residential complex, this was necessary and required. Attachment to a place is a tangible concept that fosters both social and material sustainability in urban environment. Population increase and technical advancements have made mass development and vertical residential complexes necessary. However, the affection and good feelings that residents have for their surroundings must be appreciated.

Open urban areas are therefore necessary for inhabitants' functional, social, psychological, and healthy needs, as well as for social interactions, communication, and engagement with others to take place in residential complexes. The physical and psychosocial structures are symbiotic, and the elements of the physical structure are meaningless. They can only be understood in terms of the components of the social system. Thus, place attachment is a major and tangible factor in evaluating the relationship between a person and a place, with its Psychological, social and physical aspects (Al-Mendilawi & Al-Saaidy, 2024).

Gentrification and related urban and socio-spatial transformations such as rising rents, demographic turnover, and commodification of culture, disrupt all types of attachment dimensions in two ways:

- Material/behavioral rupture that includes loss of familiar landmarks, displacement of routine sites of interaction, and closure of local businesses reduce behavioral anchors for attachment. In addition to symbolic/cognitive, rupture that is formed from commodification or "museumification" of heritage replaces lived meanings with tourist staging, undermining residents' place identity and producing feelings of estrangement (Lees et al., 2008; Zukin, 2010).
- Physical alteration of landmarks and streetscapes undermines cognitive meanings; displacement and turnover reduce daily social contacts that sustain emotional bonds; and loss of opportunities to use and care for places weakens behavioral commitments (Lees et al., 2008; Zukin, 2010). Even when buildings are preserved, the "museumification" of heritage can commodify meanings and exclude residents from practices that once constituted belonging, producing symbolic displacement (Lewicka, 2011). These disruptions help explain why psychosocial harm (loss of identity, anxiety, reduced civic efficacy) often accompanies market-led change and why interventions that restore opportunities for meaningful interaction and stewardship can help rebuild attachment.

These ruptures weaken social capital, increase psychosocial stress, and reduce adaptive collective capacity, which need to remediate via urban acupuncture.

2.2.4 Significance of Place Attachment in Heritage Contexts

In heritage contexts, strong place attachment helps sustain everyday practices, cultural transmission, and collective care—conditions that underpin social resilience—whereas its disruption by market and physical change is a primary pathway through which gentrification erodes community well-being (Manzo & Perkins, 2006; Zukin, 2010).

Within these neighborhoods, place attachment combines emotional ties to people with strong cognitive ties to history and tradition. Therefore, preserving material fabric alone is insufficient; sustaining intangible practices and everyday routines is essential to keep attachment intact (UNESCO, 2003). Interventions that restore functional uses and community agency, not only façades, are more likely to rebuild attachment and resilience (Manzo & Devine-Wright, 2013). Accordingly, urban acupuncture's small, culturally attuned interventions can be evaluated for their success in rebuilding attachment (and thus resilience) by tracking affective, cognitive and behavioral indicators.

The literature supports hypothesizing that urban acupuncture strengthens emotional, cognitive and behavioral bonds (place attachment), which in turn increase cohesion, adaptive capacity and recovery, especially in heritage contexts where identity and memory are central (Lerner, 2014; Lewicka, 2011; Norris et al., 2008).

Various considerations could be addressed concerning place attachment in heritage contexts:

1. Affective Bond (Emotional Attachment): Strength of positive feelings and emotional ties toward place such as warmth, pride and comfort (Lewicka, 2011; Scannell & Gifford, 2010). Strong affective bonds predict willingness to steward (resilience: coping/adaptation) and influence receptivity to acupuncture interventions (e.g., emotional response to a restored courtyard).
2. Cognitive-Symbolic Attachment (Place Identity & Memory): Extent to which the place is integrated into personal/group identity and narrative; includes collective memory and symbolic meanings such as place identity and historical narratives (Lewicka, 2011; Proshansky et al., 1983). Cognitive-symbolic attachment protects cultural continuity (resilience: transformability) and helps tailor acupuncture interventions to reinforce heritage narratives rather than replace them.
3. Behavioral Attachment (Place Dependence & Use): Functional, practice-based connection—the degree to which people use and depend on place for daily activities and routines (Williams & Vaske, 2003; Stokols & Shumaker, 1982). Behavioral attachment is the channel through which acupuncture's multi-functionality shows impact (increased activity is equal to increased attachment) and underpins coping capacity (resilience).
4. Social Embeddedness (Relational Attachment & Networks): Degree to which attachments are embedded in social networks—ties to neighbors, mutual support, civic participation (Manzo & Perkins, 2006; Perkins et al., 1996). Social embeddedness is critical for resilience (cohesion dimension) and influences co-design success for acupuncture projects.
5. Cultural Continuity (Intangible Heritage & Rituals): The ongoing practice, transmission, and visibility of place-specific cultural traditions, crafts, languages, and rituals (UNESCO, 2003; Zukin, 2010). Cultural continuity directly ties to identity (cognitive-symbolic) and is a crucial resilience resource (transformative capacity) that acupuncture must protect/enhance.
6. Stewardship and Agency (Behavioral Intent to Protect and Act): Residents' sense of responsibility, perceived efficacy, and actual engagement in protecting and improving place (volunteering, advocacy, care). Stewardship is a practical manifestation of place attachment that mediates between acupuncture interventions and resilience outcomes; strong stewardship predicts better project sustainability.

Place attachment refers to the emotional, cognitive, and behavioral bonds that individuals and groups develop with their physical environment, shaped through repeated interaction, memory, and shared meaning (Lewicka, 2011; Scannell & Gifford, 2010). In heritage neighborhoods, place attachment is strongly associated with everyday practices, collective rituals, and the continuity of use of culturally significant spaces. Within the scope of this research, place attachment is not measured through surveys or interviews; instead, it is assessed indirectly through observable and documented indicators such as continuity of social practices, patterns of space appropriation, local stewardship behaviors, and the preservation or reactivation of symbolic urban elements. These indicators are widely recognized in environmental psychology and

Table 2.1 Proxy indicators of place attachment

Set of proxy indicators of place attachment		
Emotional	Cognitive	Behavioral

urban studies as valid proxies for evaluating attachment where primary data collection is constrained (Lewicka, 2011; Manzo & Perkins, 2006).

In this book, place attachment operates as the psychological mediating construct or mechanism through which urban acupuncture interventions contribute to social resilience under gentrification pressure thus transforms targeted design actions into durable social resilience outcomes.

The place attachment that will serve as the foundation for study and findings are displayed in Table 2.1.

2.3 Social Resilience

Social resilience refers to a community's ability to adapt, recover, and learn from adversities (natural disasters, economic crises, etc.) It encompasses dimensions like social cohesion, adaptive capacity, and rapid recovery. This resilience could be represented in terms of multiple asset domains. For example, social capital (networks, trust), human capital (skills, education), and cultural capital (shared identity) together enable communities to solve problems and mobilize resources.

2.3.1 Historical Background of Social Resilience

The term "resilience" has acquired prominence over the past ten or more years across a variety of academic disciplines and research fields, from biology and engineering to sustainability studies, natural catastrophe research, and positive Psychology.

Resilience in positive psychology refers to the ability to cope with life hardships. Some people are knocked down by challenges, but they return as a stronger person more steadfast than before. We call these people resilient. A resilient person works through challenges by using personal resources, strengths, and other positive capacities of psychological capital like hope, optimism, and self-efficacy. Overcoming a crisis via resiliency is often described as "bouncing back" to a normal state of functioning. Being resilient is also positively associated with happiness (Pennock, 2017).

Figure 2.4 illustrates the concept of resilience as a multifaceted ability to recover from challenges by focusing on different areas of well-being. Its core components are meaning in life, positive emotions, social support, coping, and physical well-being.

Fig. 2.4 Resilience concept. *Source*: https://thriveottawa.org/resilience-2/

Originating in ecological theory, resilience has been described in two principal ways: (1) as the magnitude of disturbance a system can absorb without fundamentally altering its structure or function, and (2) as the rate at which a system returns to its original state following a perturbation (Perrings, 1998). "Crawford Holling" (1973) first used the term "resilience" in ecology, to characterize ecosystems' ability to endure disruption. According to "Holling", ecological resilience is the capacity of a system to absorb change and still maintain the same relationships among its components. This shifted thinking away from single-equilibrium "stability" towards multiple stable states and adaptive feedbacks. The idea of "Holling" emphasized diversity and non-linear dynamics, laying groundwork for later resilience thinking (Gunderson & Holling, 2002).

The foundation for the development of the concept was a basic recognition of social resilience as the ability to adapt. Later, the concepts of learning and adapting were included to form a composite definition that recognizes the importance of the roles played by politics, power, and involvement in the face of escalating uncertainty and surprise. Resilience thinking evolved to include not just the capacity to absorb shocks but also to learn, transform, and adapt, underscoring the influence of governance, power dynamics, and stakeholder participation (Folke et al., 2010).

In the 1990s–2000s, resilience thinking expanded to socio-ecological systems. Researchers defined resilience in linked human–environment systems as the capacity to absorb disturbance and reorganize while retaining core functions (Walker et al., 2004). They introduced related concepts of adaptability (managing resilience) and transformability (creating new regimes). "Folke" and colleagues (2006) further framed resilience as key to sustainability, stressing how social and ecological feedbacks allow systems to adapt or transform. Scholars such as "Adger" (2000) even equated social resilience with community-level capacity to cope with change, highlighting overlaps between social capital and ecological resilience.

By the 2000s, the resilience concept had moved into urban planning, community development and disaster recovery. Urban scholars began speaking of "resilient cities", designed to be strong yet flexible. For example, "Godschalk" (2003) describes resilient cities as "constructed to be strong and flexible" so that infrastructure (roads, utilities, etc.) can keep functioning during floods, earthquakes or other shocksresearch-legacy.arch.tamu.edu. Likewise, community resilience frameworks

Fig. 2.5 Social resilience concept. *Source*: Kwok et al. (2016)

(e.g. Norris et al., 2008) emphasize local networks and institutions that anticipate risks, limit impacts, and recover quickly. Planning and disaster programs (FEMA's initiatives, UNDRR campaigns) adopted resilience goals; advocating urban and community systems combine redundancy and adaptability (Beatley, 1998; Steinberg & Burby, 2002).

Resilience is now a transdisciplinary paradigm. In psychology, resilience traditionally referred to individuals' or families' positive adaptation in the face of adversity (Masten, 2001), but scholars have extended it to communities and cultures (Fleming & Ledogar, 2008). In sustainability science, Holling's ideas informed concepts like panarchy (Gunderson & Holling, 2002) and adaptive governance. Urban planning and public health have likewise embraced resilience as a goal of design and policy. Thus from Holling's ecological roots, resilience has evolved into a broad framework for understanding and enhancing the capacity of ecosystems, cities, and communities to absorb shocks and transform when needed (Folke et al., 2010; Walker et al., 2004).

Figure 2.5 illustrates the core components that contribute to social resilience, which is the ability of a community to withstand, adapt to, and recover from environmental, social and political changes.

2.3.2 Definitions and Dimensions of Social Resilience

All definitions of social resilience place a strong emphasis on how well social entities like persons, organizations, or communities can withstand, absorb, manage, and adapt to various sorts of environmental and social dangers (Obrist et al., 2010).

Some reliable and comprehensive definitions of resilience and social resilience are presented below:

- Resilience as the capacity of linked social–ecological systems to absorb disturbance, reorganize, and continue to develop (Walker et al., 2004).
- Resilience is the capacity of a system to deal with change and continue to develop; in social systems this includes learning, adaptive governance, and transformability (Folke, 2006).
- Community resilience is a process linking a set of networked adaptive capacities (resources with dynamic attributes) to a positive trajectory of functioning and adaptation after a disturbance (Norris et al., 2008).
- Social resilience refers to the ability of groups or communities to cope with external stresses and disturbances as a result of social, political and environmental change (Adger, 2000).
- Community resilience is the existence, development, and engagement of community resources by community members to thrive in an environment characterized by change, uncertainty, unpredictability, and surprise (Magis, 2010).

In alignment with the objectives of this book, we propose the following definition: Social resilience is "the community-level capacity to absorb, adapt to, and reorganize in response to gentrification-related disturbances—maintaining or restoring social cohesion, cultural identity (including intangible heritage), and functional community networks—through collective action, learning, and the mobilization of place-based resources".

This suggested definition explicitly connects resilience to identity and heritage, focuses on collective processes (not just individual coping), and allows for interventions (urban acupuncture) that target place attachment as a mediating mechanism.

After defining social resilience as the ability of a society to absorb, adjust to, and restructure in the face of disruption, it is helpful to break down this general concept into observable and quantifiable operational components. Drawing on resilience and community studies, three interrelated dimensions—social cohesion (networks, trust, reciprocity), adaptability (learning, flexibility, inclusive governance), and recovery (speed and completeness of restoring core functions)—capture the processes by which communities withstand shocks and rebuild (Magis, 2010; Norris et al., 2008; Walker et al., 2004). Treating resilience through these dimensions both clarifies theoretical links to place attachment and provides practical indicators for empirical assessment in heritage contexts.

1. <u>Social Cohesion</u>

Social cohesion refers to the density and quality of interpersonal ties, norms of reciprocity, and institutional trust that enable collective action (Magis, 2010; Norris et al., 2008). Cohesive communities share information rapidly, mobilize resources informally, and sustain everyday caregiving and mutual aid—capacities that both prevent cascading harms and underpin longer-term adaptation. The indicators of social cohesion are social network density, membership in local associations, frequency of neighbor-to-neighbor support, trust in local institutions, and participation rates in community events. Cohesion is tightly linked to place attachment, since emotional bonds to place foster the social ties that constitute cohesion.

2. Adaptability

Adaptability denotes the ability of a community to learn from disturbances, reconfigure practices, and modify institutions and behavior in response to changing conditions (Folke, 2006; Walker et al., 2004). It includes leadership, knowledge flows, inclusive decision-making, and flexible use of space and resources. The following are its indicators: presence of participatory governance processes, evidence of adaptive learning (post-event reflection leading to change), diversity of livelihood strategies, and existence of flexible, multi-use public spaces. Urban acupuncture can enhance adaptability by creating low-risk experimental sites where new practices are tested and community knowledge is mobilized.

3. Recovery

Recovery captures the community's capacity to absorb shock and restore core functions—housing stability, social services, economic activity –within an acceptable timeframe (Cutter et al., 2010; Norris et al., 2008). It concerns both immediate coping and medium-term restoration of social and material infrastructures. Indicators: time to restore essential services, rehousing rates, recovery of local businesses, and self-reported restoration of well-being. Strong place attachment supports recovery because residents invested in place are more likely to participate in rebuilding and stewardship activities.

Together, cohesion, adaptability, and recovery form an interdependent resilience bundle: cohesion supplies the social capital; adaptability enables systemic learning and reconfiguration; recovery measures the practical ability to re-establish function after disturbance. In heritage contexts, interventions that strengthen place attachment (through culturally sensitive urban acupuncture) can bolster all three dimensions by reactivating social spaces, enabling local learning, and anchoring post-disturbance restoration efforts (Magis, 2010; Norris et al., 2008; Walker et al., 2004).

Figure 2.6 outlines a conceptual framework for social resilience in heritage contexts, highlighting how community strength is built through interconnected dimensions: social cohesion, adaptability and recovery.

From its origins in ecology—where resilience described an ecosystem's capacity to persist under disturbance—scholars have progressively reframed the concept to foreground social transformation in the face of global change. This genealogy foregrounds three interrelated capacities central to resilience thinking: persistence (the ability to maintain core functions), adaptability (the capacity to learn and change responses), and transformability (the potential to reconfigure structures when existing regimes are untenable). Framing resilience in these terms clarifies why research must address not only recovery after shock but also longer-term processes of learning, governance, and systemic change in communities (Folke, 2006; Holling, 1973; Walker et al., 2004).

This particular genealogy is representative of the fundamental ideas that underlying the resilience notion, notably persistability, adaptability, and transformability.

Fig. 2.6 Social resilience in heritage contexts. *Source*: Authors

1. Resilience as persistability:

One of the most important studies on the subject of resilience is Hollings's "Resilience and Stability of Ecological Systems," published in 1973. With the use of instances like spruce budworm outbreaks and their effects on Canadian boreal forests, "Holling" advanced the case that ecosystems will reveal nonlinear dynamics (Holling, 1973, 1996) With his study, he substantially rebutted earlier static, equilibrium-based models of ecosystems. Instead, he advocated thinking of them as complex, adaptive systems with a variety of potential stable states, or "basins of attraction," that would perpetuate cyclicity (Walker & Salt, 2012).

According to "Holling", "the durability of ecological systems and their capacity to absorb change and disturbance and still sustain the same links between populations and state variables". This ecological resilience was measured in terms of the amount of disruption that a system could withstand while continuing to function (Carpenter et al., 2001).

2. Resilience as adaptability:

Empirical case studies on coupled social-ecological systems have been done by researchers from the Resilience Alliance, an international and interdisciplinary research network, to support the idea of resilience (Walker 2002; Carpenter et al., 2001). The "adaptive cycle" meta-theoretical model was created by fusing conceptual issues with practical observations (Holling, 2001; Gunderson & Holling, 2002; Berkes et al., 2003). The adaptive cycle, a heuristic model, shows a four-phase cyclicity of complex systems that is endogenously driven. These systems often experience the four phases of accumulation and expansion, stagnation, unexpected collapse, as well as re-organization and rejuvenation.

3. Resilience as transformability:

One of the basic pillars of resilience thinking, according to "O'Brien" and colleagues (2009), is that "environmental problems cannot be addressed in isolation from the social context." Thus, the concept of resilience might be seen as a call to action from natural scientists to social scientists to engage in integrative research with sustainability as the central idea and normative goal. In response to the criticism of conservatism expressed by critical social scientists in reaction to reading the concept of resilience as applied to social systems, resilience proponents most recently revised their definition by incorporating the idea of transformation or transformability (Pelling & Manuel-Navarrete, 2011).

Changes in internal structures, feedback loops, or exposure to a specific shock or stress may cause the system to switch from one functional basin to another. The ability of a system to change the stability landscape and create alternative system pathways in the event that ecological, economic, or social factors make the current system unsustainable is what is meant by the concept of transformability (Folke et al., 2010; Walker et al., 2004).

Table 2.2 illustrates the four-step process used to develop a broad framework for social resilience that is inclusive and flexible to different communities and disaster circumstances. Each framework's indications and traits for social resilience were listed in the first step.

Cities and other urban areas are intricate, multi-actor organisms that change across time and location. They include a variety of tangible and intangible capital assets that

Table 2.2 The three capacities of social resilience. *Source*: Authors from earlier research by Voss (2008), Lorenz (2012), Orbist et al. (2010), Béné et al. (2012) and Keck (2014)

	Coping capacities	Adaptive capacities	Transformative capacities
Response to risk		ex-post > ex-ante	ex-ante > ex-ante
Temporal scope		short-term	long-term
Degree of change		low, status quo	high, radical change
Outcome		restoration of present level of well-being	security of future well-being
Outcome	restoration of present level of well-being	security of future well-being security of future well-being	enhancement of present and future well-being

Table 2.3 Set of parameters of social resilience

Set of parameters of social resilience		
Coping capacities	Adaptive capacities	Transformative capacities

can move both slowly and quickly (Andersson & Andersson, 2017; Kourtit et al., 2022).

The social resilience parameters that will serve as the foundation for study and findings are displayed in Table 2.3.

Social resilience operates simultaneously at individual, household, neighborhood, and institutional scales that should be matched to the scale of interest (e.g., individual well-being vs. neighborhood cohesion). We have to note that resilience is not neutral: power relations shape who benefits from adaptation or transformation. Measurement must therefore include equity indicators such as vulnerability and access to decision-making (Adger, 2000). Social capital (trust, reciprocity) and place attachment function as mechanisms that link interventions (urban acupuncture) to resilience outcomes, and so require both attitudinal and behavioral measures. It is worthy to note that some resilience goals (e.g., preserving a particular cultural identity) are normative choices; measurement frameworks should therefore be transparent about values and stakeholder priorities.

2.3.3 Social Resilience in Architecture and Urban Design

Social resilience describes a community's capacity to cope with, adapt to, and transform in response to shocks and stresses while maintaining essential social functions and identity. In psychological—social—ecological systems literature, this is often framed as three temporally nested capacities: coping (short-term buffering), adaptive (learning and adjustment), and transformative (system-level change). These concepts have direct implications for architecture and urban design because the material and spatial environment both shapes and is shaped by community behaviors, practices, and governance arrangements (Folke, 2006; Magis, 2010; Walker et al., 2004).

Urban design contributes to social resilience not only by providing hazard – resistant infrastructure, but also by creating conditions for everyday social life—the routines, meanings and networks that enable communities to respond collectively to change. Urban resilience must be understood across social, infrastructural and ecological domains; designers operate at the intersection of these domains by shaping the places where social capital, informal economies, and civic practices are reproduced (Meerow et al., 2016).

The four phases of "exploitation," "conservation," "release," and "reorganization" are continuously experienced by urban systems, which are dynamic entities with non-equilibrium dynamics. In order to deal with the complexities and dynamics of urban systems, resilience-oriented planning is required (Sharifi & Yamagata, 2018).

Four casual pathways explain why architecture matters in social resilience:

1. Facilitating daily routines (behavioral capacity): Accessible, cozy, and secure public areas promote frequent use (walking, meeting, informal commerce), which creates routines and mutual acquaintanceship—the "social glue" that facilitates quick mutual aid in emergencies. Continuous sidewalks, benches, distinct sight-lines, lighting, and mixed-use façade are examples of design elements (Norman, 2013; Scannell & Gifford, 2010).
2. Reinforcing shared meanings (cognitive capacity): Material cues, heritage markers, and interpretative design uphold communal memory and identity, strengthening locals' dedication to the area and willingness to contribute to its restoration and maintenance. Preserved façades, plaques, memory nodes, and locally significant materials are examples of design elements (Lewicka, 2011; Scannell & Gifford, 2010).
3. Encouraging adaptable behaviors (adaptive capacity). Communities can test new uses and modify routines without permanently upsetting the cultural fabric thanks to adaptable, multipurpose venues and progressive, reversible interventions. Pop-up markets, modular furniture, and adaptable lighting and programming are design elements (Ahern, 2011; Till & Schneider, 2005).
4. Using transformative potential to catalyze institutional change Alternative management or governance models (community stewardship, micro-enterprise support) that scale up into policy changes can be demonstrated by planned and strategically positioned initiatives. Civic platforms, community-run kiosks, and pilot-to-policy paths are examples of design elements (Walker et al., 2004; Vale & Campanella, 2005).

Accordingly, certain design approaches consistently support social resilience in the research and in practice:

- Multi-functionality: create places that host several activities across time to sustain daily activation (Brandt & Vejre, 2004).
- Flexibility & reversibility: prefer low-cost, demountable solutions for testing and iteration (Till & Schneider, 2005).
- Legibility & predictability: clear movement hierarchies and wayfinding reduce cognitive load and encourage routine behavior (Norman, 2013).
- Embedded cultural cues: integrate heritage materials, narrative devices and symbolic elements to protect cognitive attachment (Lewicka, 2011).
- Participatory programming: co-design and community stewardship programs build ownership and support adaptive governance (Magis, 2010).
- Monitoring & low-barrier metrics: simple indicators (pedestrian counts, event tallies, photographic transects) allow designers to iterate and demonstrate social benefits.

Below are tangible examples of how design and planning contributed to social resilience in various scales and in different socio-political contexts.

Example 1: City of Medellín, Colombia (2004)

Years of social breakdown and violent crime plagued the Colombian city of Medellín. Residents of the city's most remote and impoverished hillside neighborhoods used to have a 2.5 h walk into the city. In contrast, a new mayor decided in 2004 that every city investment should have a social component, and within a year they had ordered and delivered a new cable car transit system, which is a model of social cohesion in action. All of a sudden, the city's centre was only 20 min away, there were more job opportunities, and crime had decreased. The cable car project has evolved into a brilliant and concrete illustration of how to use technology to foster shared urban experiences and increase social mobility.

Medellín's municipal strategy used public transport, libraries, cable cars, and public spaces to reconnect marginalized barrios to the urban core, intentionally combining social programs with spatial upgrades. The interventions strengthened daily mobility, civic identity and opportunities for collective action—contributing to declines in violence and increased civic participation. Medellín demonstrates how coordinated, multi-layered design can enhance coping and adaptive capacities when coupled with governance commitment (Guerra, 2014).

Figures 2.7 and 2.8 display the Metro cable system in Medellín, Colombia, an innovative aerial cable car transport network. It serves as public transit solution, designed to improve accessibility and social inclusion, fostering resilience and connecting diverse communities.

Example 2: Rosenthal Park in Sydney, Australia (2014)

The Rosenthal Park displays a landscape that attends to the community's social, economic, environmental, and cultural needs by enhancing the integration and

Figs. 2.7 & 2.8 Cable cars system. *Source*: Taipai times, https://www.taipeitimes.com/

connections with the resilience concept in public spaces. It serves as both a public space and a neighborhood backyard.

In Fig. 2.9, one could divide the Rosenthal Park into two sections. The first section is about open spaces that offer opportunities for a variety of events, like a central location for flexible gathering, dispersed play areas for children of all ages, and performance areas for a variety of community activities. Due to this feature, the park is incredibly adaptable and diverse, allowing visitors to interact with both the surrounding natural beauty and one another while enjoying a variety of social gatherings and recreational activities. Facilities that are sustainable and interactive make up the park's second section. The environmentally friendly facilities include native wildflower planting beds, solar energy collection structures, rainwater disposal systems, and roof gardens.

Figure 2.9 displays the landscape master plan for the redevelopment of a council-owned site into a public space, identified as the Lane Cove Town Centre in Australia. The plan includes a multi-level design with retail space and three levels of underground parking, featuring a boardwalk, active event spaces, and natural landscape elements. It is designed to enhance community engagement through nature and water play areas.

Example 3: Sankt Kjeld' Square & Bryggervangen in Copenhagen, Denmark (2019)

Sankt Kjeld's Square form a cornerstone in the Copenhagen Climate District and is one of the largest and greenest cloudburst mitigations projects to date. The project shows how making our neighborhoods resilient to future cloudbursts can go hand in

Fig. 2.9 The layout and sections of the Rosenthal Park in the Lane Cove Community. *Source*: World Landscape Architecture, https://worldlandscapearchitect.com/

Figs. 2.10 & 2.11 The Sankt Kjeld' Square & Bryggervangen Community in Copenhagen. *Source*: https://www.sla.dk/cases/sankt-kjelds-square-and-bryggervangen/

hand with green and recreational urban spaces that enhance biodiversity and reduce traffic, air pollution, and the urban heat island effect.

In response to climate change strengthening resilience to flood invasion through careful planning of the community's physical layout. The Skt. Kjeld Community has made the best use of open spaces by constructing a number of number of green spaces that are structurally connected and operationally interconnected, such as green island plazas, cloudburst boulevards, green courtyards, and public areas, as shown in Figs. 2.10 and 2.11. This project develops a robust spatial network that offers more opportunities for nature-based cloudburst protection combined with recreational spaces, biodiversity, and new infrastructure.

Figures 2.10 and 2.11 show (Sankt Kjeld's Square) in Copenhagen, Denmark, a notable example of sustainable urban planning and landscape architecture. The square has been converted from a congested intersection into a vibrant green public space that serves as a climate-adaptive "sponge city," addressing heavy rainfall and flood prevention by functioning as a storm water reservoir. The project has more than doubled green space, enhancing urban biodiversity with a variety of plants and trees.

2.3.4 Significance of Social Resilience in Heritage Contexts

Social resilience has emerged as a critical concept in architectural and urban discourse, particularly within heritage contexts where physical fabric, cultural meaning, and community life are simultaneously exposed to disruption. Unlike purely material approaches to conservation, social resilience foregrounds the capacity of communities to sustain identity, continuity, and collective functioning while adapting to economic, social, or environmental change (Adger, 2000; Vale & Campanella, 2005).

In heritage neighborhoods, preservation is not solely a matter of safeguarding buildings, but of maintaining the social practices, memories, and symbolic meanings that give those environments life. Social resilience enables communities to remain active stewards of their heritage, reinforcing everyday use, informal care, and intergenerational transmission of place meanings (Lewicka, 2011; Pendlebury, 2013). When residents possess strong coping and adaptive capacities, heritage spaces are more likely to be inhabited, maintained, and defended against erasure or commodification. Thus, socially resilient communities function as living conservation agents, ensuring that heritage remains embedded in daily life rather than reduced to static or touristic artifacts.

Heritage districts are particularly vulnerable to disruption caused by gentrification, tourism pressure, post-conflict reconstruction, or environmental stress. These forces often destabilize social networks, displace long-standing residents, and fracture place attachment (Smith, 2006; Zukin, 2010). Social resilience provides a framework to understand how communities absorb such shocks (coping), reorganize spatial and social practices (adaptation), and, in some cases, reconfigure governance or spatial structures to secure long-term continuity (transformation) (Meerow et al., 2016).

Case studies from post-war cities (e.g., Baghdad, Beirut), historic centers under tourism pressure (e.g., Cairo, Venice), and informal heritage settlements illustrate that resilience is rarely achieved through large-scale redevelopment alone. Instead, it often emerges through incremental, community-centered interventions that reinforce social ties, everyday routines, and collective agency—precisely the scale at which urban acupuncture operates (Hamdi, 2010; Vale, 2014).

From an architectural and urban design perspective, social resilience shifts attention from form to process, emphasizing how spatial interventions enable social interaction, mutual support, and adaptive use. Design strategies that enhance legibility, accessibility, multi-functionality, and cultural recognition contribute directly to resilient outcomes by strengthening residents' sense of control, predictability, and belonging (Ahern, 2011; Norris et al., 2008). In heritage contexts, this means that resilience is not achieved by resisting change entirely, but by guiding transformation in ways that preserve social continuity and local identity.

Accordingly, social resilience serves in this research as a key evaluative outcome, capturing how communities respond to gentrification pressures and how design-led interventions can either erode or reinforce the social foundations of heritage environments.

Figure 2.12 operates a conceptual diagram that outlines six key resilience attributes for historic neighborhoods—from knowledge/skills and community networks to "people–place connections" (sense of belonging)—each with associated indicators. For example, it shows how cultural heritage can support local knowledge and capacity building and how a strong identity and attachment to place ('people–place connections') foster collective resilience. The framework highlights that strong social bonds and shared cultural narratives (place attachment) underlie adaptive capacity in

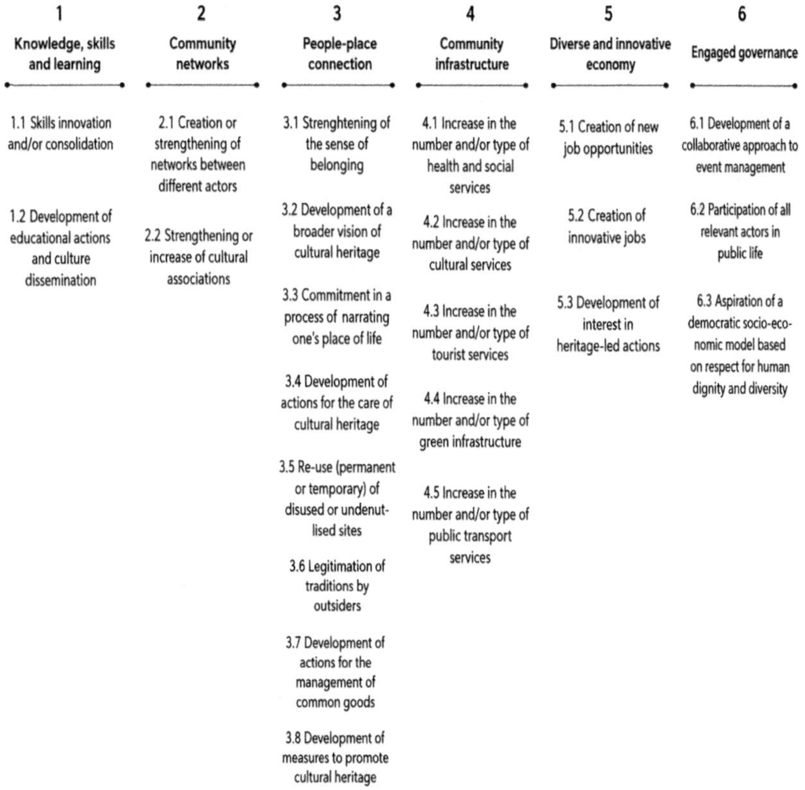

Fig. 2.12 Heritage community resilience attributes and indicators. *Source*: Fabbricatti et al. (2020)

heritage communities, reinforcing social cohesion and well-being (Fabbricatti et al., 2020).

Smaller, targeted interventions such as parklets, pocket parks, improved crossings, lighting and pop-up markets—often produce measurable increases in pedestrian use, nighttime safety perceptions and local commerce. Such projects (e.g., parklet programs in many cities, or the Kodak Passageway and LibreLigne examples in this book) show how micro-scale design can deliver immediate coping benefits and, when sustained, catalyze attachment and stewardship (Lerner, 2014).

Design alone does not guarantee resilient social outcomes. The evidence shows that design is necessary but not sufficient: it works best when coupled with enabling policies (anti-displacement measures, affordable tenure), supportive institutions (maintenance budgets, inclusive governance), and culturally sensitive programming. In some cases, even well intentioned upgrades have accelerated rent increases and displacement; a reminder that architects and designers must consider socio-economic effects and embed equity safeguards from project inception (literature on gentrification and displacement).

2.4 Urban Acupuncture: Effective Interventions for Heritage Contexts

Urban acupuncture is an urban design strategy that targets small, strategic "pressure points" in a city (often underused or overlooked spaces) with low-cost, community-driven interventions to produce wider positive effects. Just as acupuncture needles stimulate overall health by acting on critical points, urban acupuncture projects like (parklets, community gardens or repurposed structures) aim to revitalize neighborhoods by improving local quality of life. This approach explicitly views cities as "living organisms" and uses pinpointed interventions as "needles" to heal specific urban ills (Casagrande, 2010).

Urban acupuncture emphasizes "pinpoint" interventions at strategic urban nodes; it frames interventions as "pinpricks" in the urban fabric, precise, site-specific acts intended to reactivate circulation, social use and symbolic meaning. Thus, urban acupuncture forms a pragmatic design approach for activating social life in contested heritage quarters.

2.4.1 Concept and Origins of Urban Acupuncture

Acupuncture, a form of traditional Chinese medicine, is thought to hasten the recovery process for a variety of physical and mental conditions. Drawing from the metaphor of the Chinese traditional medical practice, urban acupuncture refers to the improvement of social and urban issues through precise interventions that revitalize areas of the city, and consolidate urban planning strategies.

The term urban acupuncture traces back to the Spanish architect "Manuel de Solà-Morales" (1999–2004) and gained prominence through "Jaime Lerner's" work (2003–2014) and related theorists. The concept circumscribes projects with a high degree of reversibility, allowing for corrective measures and improvements. The tactic has the advantage of being a fast-tracked planning, thus being implemented swiftly and often with modest means. These localized actions can start in a chain reaction that improves the quality of the surrounding urban environment, from public safety to social cohesion, with fewer bureaucratic requirements.

Urban acupuncture is an innovative applied solution for revitalizing urban fabric. In the urban acupuncture approach, the targeted area is the key point within the urban environment. This point is not energetic enough due to some reasons, but it is important to its surroundings. A low-cost and low-effort intervention makes this area work better and last longer. Urban acupuncture, by using small-scale simulation, pushes relief into the small site and transforms the energy from it to the larger context. Urban acupuncture, as described by "Morales", is an idealistic approach for the current limited economic era, it is "an idealistic approach for the current economic era". "Lerner" demonstrated the necessity of community involvement in the practice of urban acupuncture: "Just as good medicine depends on the interaction between

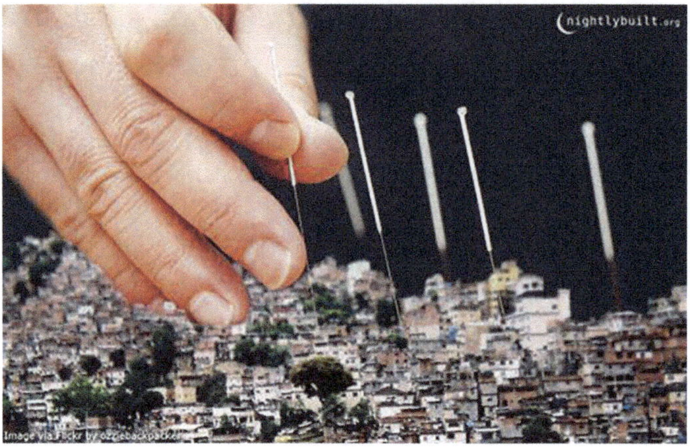

Fig. 2.13 Urban needles. *Source*: https://alchetron.com/

doctor and patient, successful urban planning involves triggering healthy responses within the city, probing here and there to stimulate improvements and positive chain reactions. Intervention is all about revitalization, an indispensable way of making an organism function and change" (Dikmak et al., 2024).

Figure 2.13 illustrates the concept of "Urban Acupuncture," a metaphor used in urban design and planning to describe small-scale interventions intended to revitalize a larger urban area. The metaphorical concept likens targeted urban interventions to medical acupuncture, proposing that small changes in neglected areas can lead to broader positive transformations in neighborhoods. Developed by architect "Jaime Lerner", this theory emphasizes quick, impactful modifications over extensive master planning and is typically applied in favelas or underserved waterfronts to enhance social and environmental conditions.

Three times mayor of Curitiba, architect and urban planner "Jaime Lerner" is one of the lead advocates of urban acupuncture, seeing it as a means to bring immediate improvements to the urban environment, bypassing long decision—making processes and surpassing economic impediments. As he sees it, "the lack of resources is no longer an excuse not to act. The idea that action should only be taken after all the answers and the resources have been found is a sure recipe for paralysis". It is worth mentioning at this point that urban acupuncture is a multi-scalar tactic, with projects varying in size, from a reorganization of a corner street to the implementation of new transport lines.

Figure 2.14 presents the cover of the book "Urban Acupuncture" by "Jaime Lerner", which celebrates small-scale urban projects—"pinpricks" of change—that can significantly enhance city life. "Lerner", a renowned architect and former mayor of Curitiba, Brazil, pioneered sustainable urbanism. The book emphasizes how minor, targeted interventions can serve as catalysts for positive change and community revitalization.

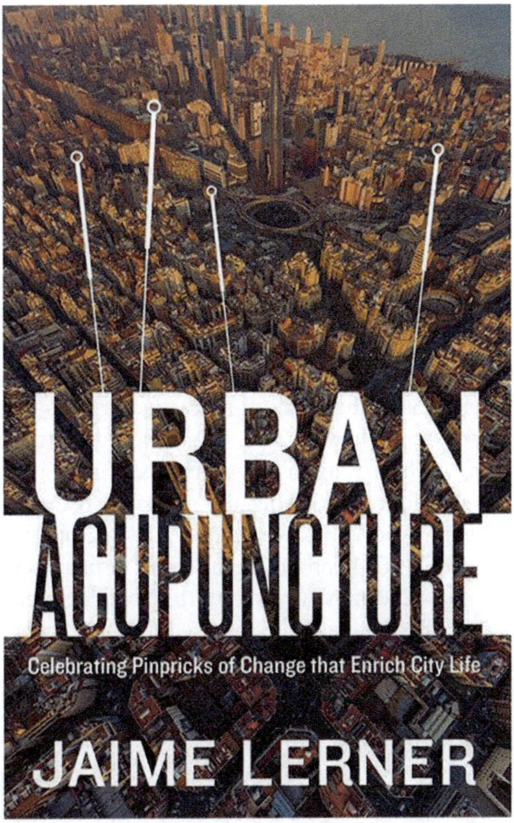

Fig. 2.14 Urban acupuncture as "pinpricks" of urbanism—small interventions that ripple outward to produce systemic change - Lerner's book summary page. *Source*: Lerner, J. (2014). *Urban Acupuncture*. Island Press

Urban acupuncture, popularized by "Lerner" and others, draws analogy to medical acupuncture: it uses targeted "pinpricks" or focal interventions to relieve stress in city life. The conceptual origins of urban acupuncture lie at the intersection of urban design, landscape urbanism and place making. "Lerner" presented the term in contemporary planning discourse, drawing analogies to medical acupuncture: modest, strategically placed stimulations produce systemic effects. For example, he defines it as "focal, quick, and precise interventions that help to enhance the health of all urban systems," pinpointing areas of stress to be alleviated.

"Marco Casagrande" and "David West" are two widely cited advocates of the urban-acupuncture idea, but they approach it from complementary angles. The Finnish architect "Casagrande" frames the city as a "multi-dimensional sensitive energy-organism" in which local places function like nodes of life and stress; his bio urbanism locates hidden "hot spots" of social or environmental tension and treats small, place-specific interventions as pressure-point remedies that can relieve and re-balance the urban organism.

According to "Casagrande", cities represent a complex of overlapping layers of energy that determine community behaviour (Casagrande, 2019).

"West" likewise emphasizes the minimal, tactical nature of this approach: in his 2011 commentary he described urban acupuncture as a practice of "small, subtle and light interventions" that channel local community energy to solve problems, offering a low-cost, bottom-up alternative to grand, top-down redevelopment schemes (West, 2011).

Taken together, these formulations locate urban acupuncture at the intersection of ecological metaphor and tactical urbanism: "Casagrande" supplies the bio urban rationale (the city as living system where targeted inputs produce system-wide ripples), while "West" and other practitioners articulate the procedural ethos (iterative, community-centered, and intentionally modest interventions). Both strands stress that the value of acupuncture is not simply physical repair but social activation—creating conditions for local agency, everyday use, and incremental governance change rather than one-off aesthetic upgrades (Casagrande, 2019).

As shown in Fig. 2.15, it is believed that the insertion of tiny needles into the skin will have long-lasting positive effects. Cities, like the human body, can benefit from this old adage: when they are sick, they need to be healed, according to "Jaime Lerner", who will leave behind a lasting legacy after passing away in May 2021. "Lerner" (2014) proposed the term "urban acupuncture" for some of the medical "magic" that may and should be applied in cities for sustainability and healing. Urban problems can be resolved through "pressure spots," with beneficial partial ripple effects that benefit the entire city.

(A):Urban acupuncture network by Marco Casagrande (source: Googleimage,helsinkiacupuncture.blogspot.com)

(B): Pocket Parks by Jolma Architects (source: Google image, pocket-parks-as-urban-acupuncture.com)

Fig. 2.15 Urban acupuncture strategy. *Source*: Al-Hinkawi and Al-Saadi (2020)

Figure 2.15 features an urban acupuncture network by "Marco Casagrande", conceptualizing the city as a living energy-organism, and showcases "Pocket Parks" by Jolma Architects, demonstrating practical applications of this theory to enhance urban health and well-being. The approach emphasizes strategic healing, using small city components as crucial points to incrementally elevate the entire urban landscape.

Urban acupuncture is a concept put out by a number of urban pioneers and mixes acupuncture with urban planning. According to "Casagrande", cities are intricate energy beings, with several, overlapping layers of energy dictating not only the city's growth but also the lives and behaviors of its citizens, in addition to the development of the city. With many ways of mixing the environment with urban design, "Casagrande" developed methods to manipulate the timelines of urban energy flows in order to create an environmentally sustainable urban development, what is called a "third-generation city" (postindustrial city). In this urban acupuncture, small-scale stimulating interventions transform the energy of small sites in the larger urban context. The sites are chosen through a comprehensive analysis of social, economic and environmental factors, developed in dialogue between designers and society to reduce stress in the environment through these small interventions in the urban fabric (Al-Hinkawi & Al-Saadi, 2020).

2.4.2 Principles and Significance of Urban Acupuncture

Urban acupuncture is usually strategic and theory-driven: sites are chosen through careful analysis of spatial, social, economic and ecological "energy" factors, and interventions are designed in dialogue with community stakeholders. Thus, it is considered as rapid, creative, place-based action and value community engagement.

This approach privilege human-scale change over large-scale masterplans and foreground the social processes of activation: it seeks to re-establish everyday practices, create opportunities for serendipitous encounters, and provide visible signals of care that can rebuild trust and belonging. In heritage contexts, these micro-interventions must operate with sensitivity to cultural continuity, ensuring that interventions reinforce (rather than commodify) local meanings embedded in material and intangible heritage (UNESCO, 2011; Zukin, 2010).

"Casagrande" (2012) describes urban acupuncture as a cross-architectural manipulation of the city's collective sensory thought. The city is seen as a sensitive, multi-dimensional organism, a living environment. By handling blockages and pushing relief energy around this body, it can be more responsive and sensitive to society's needs than traditional institutional forms and large-scale urban renewal interventions (Casagrande, 2012).

As "West" (2011), described Urban Acupuncture, the strategy focuses on small, subtle and light interventions that employ and positively direct community energy, including active citizens to address urban problems and improve the city's landscape. It is intended to substitute for large, top-down interventions that usually require massive investment from scarce municipal funds. More broadly, acupuncture sites in urban areas can be seen as a contact with the city outside, a natural sign of life in a city programmed to accommodate it.

"Morales" (2004) devised urban acupuncture as a small intervention strategy, a more local and social approach in an era of limited budgets and limited resources that democratically and cheaply increases the comfort of city dwellers through the corrective potential of compact catalytic urban interventions. Achievable within a somewhat short period, these small interventions can restructure their direct surroundings. Cities are living, breathing organisms, with all parts completely intertwined; sustainable projects—which serve as needles—can refresh the whole by healing the parts (Morales, 2004).

"Lerner" (2014) suggested the term urban acupuncture for some of the medical "magic" that can and should be applied to cities for recovery and sustainability. Urban problems can be addressed through "pressure points", with partial ripple effects that positively affect an entire city. At the physical, sensory level, acupuncture interventions include infrastructure, such as transportation, as in the development of Cortes, Brazil, or the addition of a building at historical sites, as in the intervention of the Louvre Museum, or a natural view, as in the Park Guell project, or using simpler treatments, as in the development project in Luanda, Angola. These interventions can achieve several goals, the most important of which are the increase in external spaces to make the city stimulate interactions among people. The more effectively functions are integrated into spaces, the more effective they will be (Lerner, 2014), as shown in Fig. 2.13.

Figure 2.16 provides a collection of examples regarding urban planning and development, specifically highlighting concepts of urban acupuncture.

- (A) Transportation network map for Curitiba, Brazil.
- (B) Park Güell in Barcelona, Spain, used as an example of sharing city spaces.
- (C) The Pyramid at the Louvre Museum in Paris, France.
- (D) Design proposal for the Bossa Nova Park in Rio de Janeiro, Brazil.
- (E) Central Park area in Santiago de los Caballeros, Dominican Republic.
- (F) The High Line Park in New York City, USA.

"Marchall" (2004) described urban acupuncture as a metastatic and strategic urban intervention for reconstruction, primarily supported by the design of public spaces, because urban acupuncture is more immediately effective in achieving goals related to public space. The term "metastatic" describes a series of measures that could be the focus of strategic renewal and act as a vital nerve affecting the center of a neighborhood and the city. These measures would be built for improving what is already there, to convert, modify, rehabilitate, confirm or create identities that are clearer and more important. The local intervention strategy makes it easier for the designer and critic to achieve their best performance at one point in time, in this way to generate new flows of energy in the city, not only in terms of the specific location, but also in relation to future developments arising from the site in ways that cannot be expected. Marchall (2024)

(A): Transport development in Curitiba.(Source:googleimagevimeo.com/107187044)

(B): Park Guell, Barcelona / Share the City.(Source google image - dirt.asla.org/2014/09/18/jaime-lerners-urban-acupuncture)

(C): Interfering of the Louvre Museum in Paris.(Source google imagedirt.asla.org/2014/09/18/jaime-lerners-urban-acupuncture)

(D): Rio de janeiro Bossa Nova Park. (Source:googleimagevimeo.com/107187044)

(E): Santiago de los Caballero Central Park. (Source: google image vimeo.com/107187044)

(F): High Line Park – New York City.(Source:googleimagevimeo.com/107187044)

Fig. 2.16 Urban acupuncture levels according to "Jaime Lerner". *Source*: Al-Hinkawi and Al-Saadi (2020)

"Foth" (2008) explained urban acupuncture, the strategy aims to relieve the "body" or "the built environment" according to the basic principles of urban environmental and social theories, namely through strategic healing, where parts of the city operate as a heart that gradually heal the whole. A small expansion of the targeted range of interventions gradually turns into the larger urban context. That is, the specific points at which to apply pressure to the object in order to manage or treat diseases or any kind of pain are determined by the areas that need to be reformed. Sites are chosen after environmental, economic and social analysis of the broader context through interaction with the community. Similar to the body's healing action, small-scale interventions create a social stimulus in the urban fabric, releasing energy flows by responding to local needs. In other words, the whole city's revival is activated by the treatment and recovery of component parts at strategic points (Marcus, 2009).

The strategy of urban acupuncture can be identified as a small-scale stimulus or intervention in areas that need to be fixed (obstruction points), functioning as a series of measures that can be the focus of strategic renewal and that shift their energy into the larger urban context. Sites are selected through a comprehensive analysis of social, economic and environmental factors, developed through dialogue between designers and society to relieve stress in the environment. This would substitute for top-down, large-scale interventions that usually require massive investment of funds, which many cities do not have at present. The intervention occurs at two levels: physical and sensory. The levels of intervention include the transportation system, adding a building or adding a natural view, as well as simpler treatments.

Most Urban Acupuncture studies indicate eight principles of applying this strategy as shown in Fig. 2.17. The first step is to determine the flow of the built environment. Then define the sick points where there is low energy or energy blockage. The final step is to formulate a scenario for each sick point. This scenario must be small but effective and act immediately with the involvement of citizens just like the acupuncture needle.

Figure 2.17 presents an octagonal diagram of eight principles for urban acupuncture, labeled from UA.P 01 to UA.P 08. It includes strategic approaches such as addressing "Sensitive spot" areas, implementing "Small-scale" projects, and taking "Quick action." The framework underscores the significance of "Scenario" planning and a "Holistic approach" for comprehensive problem solving, as well as "Citizen's participation" and "Citizens education" for community engagement. Lastly, it emphasizes "Create places" as the tangible outcome of these strategies.

1- Sensitive Spot Selection. Definition of Sick Points.

"Morales", in his book "A Matter of Things", mentioned the urban acupuncture theory by describing that the city possesses skin; any operation on the skin of the city must be precisely chosen at a point that gives energy to the city. In other words,

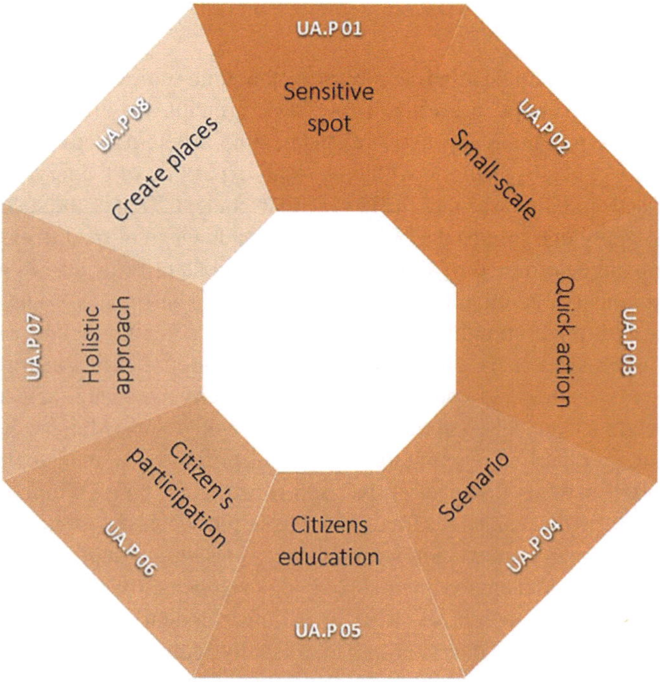

Fig. 2.17 Principles of urban acupuncture. *Source*: Dikmak et al. (2024)

"the location of the sensitive point is the first step in the strategic treatment of the urban skin". It involves discovering diseased points or points of urban blockage after considering the analysis of the built environment.

2- Small-scale intervention.

Urban Acupuncture aims for big impacts through little interventions. Morales stated that "small" does not refer to the size only: it reflects other factors, including funding, cost, and investment. Lerner also stated, "Sometimes, urban planning is just too slow and laborious," acupuncture is, therefore, "a way of supplying the city with a quick dose of energy, using few resources".

3- Quick action plan. Immediate act.

As opposed to the traditional strategies that might take time, urban acupuncture requires a quick act, quick actions, and immediate response.

4- Screenplay. Having a scenario.

According to Lerner, to achieve good practice, there is a need for a scenario and design "that appeals to the majority and creates a commitment". The principle of the need for a scenario is closely related to the degree of citizens' participation.

5- Citizen education.

People must be educated to understand the nature of their society and environment. This understanding of urban intervention accelerates the possibility of success.

6- Citizen participation and involvement.

Nowadays, people have their useful points of view, so they must be involved in the decision-making process of their community. "Lerner" says: "We cannot be so proponent on having all the answers. It is important to start and have the contribution of people; they could teach you if you are not on the right track". Citizen participation leads the experts to a completely new vision of development, as they do not have the right answers for every problem.

7- Holistic approach.

To solve the problems, all the urban aspects must be included in the interventions integrating economic, environmental, infrastructure, historical, and political elements.

8- Space into place. Creating places.

Urban acupuncture works where there is energy blockage. Some might therefore, not appreciate these places. After an intervention, the energy flow changed into a positive flow (Dikmak, et al., 2024; Lerner, 2014).

The essential principles of urban acupuncture emphasize precision, minimal intervention, and ripple effects. Interventions are precise: they resemble the medical metaphor of a needle—a "pinprick" to trigger healing. They are minimal: deliberately small in scale (often a street bench, mini-park, pop-up event) and reversible if outcomes prove unsatisfactory. They aim for a ripple effect: improving one spot should "trigger a chain reaction of improvement" in the surrounding fabric, boosting social cohesion, safety, and vitality. The strategy of urban acupuncture is characterized by:

- Precision and Targeting: Rather than large-scale, disruptive urban redevelopment, interventions are highly focused. They target specific "pressure points"—neglected, underutilized, or distressed areas—to release energy and trigger revitalization.
- Minimal Intervention (Small Scale): Interventions are deliberate, small-scale, and, often, temporary (e.g., pop-up events, pocket parks, street furniture).
- Reversibility: A key feature of urban acupuncture is that it allows for, low-risk experimentation. If a project proves unsuccessful or if needs change, the intervention can be modified or removed.
- Ripple Effects: A small-scale intervention acts as a "needle" that triggers a positive chain reaction, stimulating social, economic, and environmental improvements in the surrounding area.
- Contextual Sensitivity: Interventions are tailored to the local culture, history, and social dynamics of a specific site.
- Community Participation: Projects often emerge from, and are supported by, local residents, fostering a sense of ownership, social cohesion, and improved quality of life (Balika et al., 2021).

In practice, this might mean planting trees in a neglected alley to attract more pedestrians, restoring original heritage light fixtures to enhance a square's character, or adding temporary gardens in unused lots—each small change enhances the existing environment and signals that care is being taken. Such interventions align with urban regeneration theory by providing near-term, community-driven upgrades that build momentum for broader revitalization. Recent analyses even characterize urban acupuncture as a form of sustainable social innovation: it bridges physical regeneration and collaborative practices, generating social, economic, and environmental benefits at the neighborhood level while igniting transformation at the city scale (Casprini et al., 2026).

In summary, urban acupuncture reflects a resilience mindset: incremental and adaptive changes that empower local actors and enhance the capacity of historic urban areas to thrive in the face of resource constraints and change. The core principles of urban acupuncture are:

(a) Precision: interventions at selected nodes with high catalytic potential
(b) Scalability & replicability: small investments that can be repeated
(c) Low cost and speed: affordability and rapid deployment
(d) Participation: co-design with residents to ensure cultural fit
(e) Iterative evaluation: learning from outcomes and adapting (Lerner, 2014; Lydon & Garcia, 2015).

These principles align with theories of human-scale urbanism (Gehl) and place making (PPS), which demonstrate that well-designed small places generate disproportionate social returns by enabling everyday interactions and stewardship (Gehl, 2010).

2.4.3 Evidence of Impact (Case Studies Examples)

Urban Acupuncture is an urban environmentalism theory, which combines urban design with the traditional Chinese medical practice of acupuncture. Acupuncture relieves stress in the body; urban acupuncture relieves stress in the environment. As a design methodology, it uses tactical, small-scale interventions within the existing urban fabric, aiming to create ripple effects and transformation on the larger urban organism. Sites for Urban Acupuncture are selected through an aggregate analysis of social, economic, and ecological factors, and developed through a dialogue between designers and the community. Through the acupuncture points, Urban Acupuncture seeks to be in contact with the site-specific Local Knowledge. This strategy views cities as living, breathing organisms and pinpoints areas in need of repair. Sustainable projects, then, serve as needles that revitalize the whole by healing the parts. Urban acupuncture produces small-scale but socially catalytic interventions into the urban fabric (Casagrande, 2010).

Urban acupuncture is a design tactic promoting urban regeneration at a local level, supporting the idea that interventions in public space do not need to be ample and expensive to have a transformative impact. An alternative to conventional development processes, urban acupuncture represents an adaptable framework for urban renewal, where highly focused and targeted initiatives help regenerate neglected spaces, incrementally deploy urban strategies, or consolidate the social infrastructure of a city.

However, for this section, localized interventions, though small scale, can have a significant positive impact on their surroundings, as emphasized by the case studies focused on evidence of their effects.

Urban Bloom, a former parking lot has been transformed into an urban garden project prioritizing people. It features globe-shaped structures covered in colorful foliage, creating a vibrant atmosphere. Rich flowers and greenery are planted among the structures, promising an inviting garden as they grow. Located on Anfu Road, which merges residential and commercial spaces uniquely in Shanghai, the project highlights that urban development can thrive without demolishing the old city. This initiative addresses the lack of parks in the area, fostering community pleasure and happiness. Visitors are encouraged to engage with the space, interpreting its purpose through personal experiences. The dynamic nature of the environment and visitor behaviors shape its character, resembling the flow of a city. Urban Bloom aims to minimize waste and interact with nature sustainably, addressing urbanism's challenges of waste and consumption. Constructed from repurposed and recycled materials, it features a landscape created from wooden pallets, offering various urban settings for gatherings and events.

Figure 2.18 presents the "Urban Bloom" project is a temporary urban garden installation in Shanghai, designed by AIM Architecture and URBAN MATTERS. It transforms a former parking lot into a dynamic garden using artificial and recycled materials. The project's goal is to combat the lack of recreational spaces in rapidly developing cities, offering a landscape for gatherings, lectures, and outdoor activities. It emphasizes joy and freedom, reminding citizens of the importance of urban enjoyment and proposing a new balance between natural and fabricated environments.

Casa do Quarteirão is an initiative from the Walk&Talk 2016 arts festival, aiming to create a communal space in the neighborhood of Quarteirão for collaborative activities. It held annually in the Azores Islands since 2011, focuses on site-specific cultural projects, fostering contemporary artistic experimentation in dialogue with the local community and culture. This project is a user-driven program designed as an open project, featuring a modular frame building system that allows for adaptation to various community needs and activities. The structures, constructed in five days with residents, utilized wood from the endemic conifer Cryptomeria japonica found on the island.

Fig. 2.18 Urban Bloom pop-up garden installation in Shanghai, China (2018). *Source*: Wong, Joanna. https://www.archdaily.cl/cl/893837/urban-bloom-aim-architecture-plus-urban-matters

Fig. 2.19 Casa do Quarteirão – Portugal (2016). *Source*: Soares, Rui. https://divisare.com/projects/332350-orizzontale-rui-soares-casa-do-quarteirao

Figure 2.19 displays The "Casa do Quarteirão" installation, created by the architectural collective Orizzontale, is a public art project located in a neglected alleyway of the O Quarteirão neighborhood on São Miguel Island, Portugal, as part of the Walk&Talk Azores 2016 festival. Its purpose is to transform the space into a friendly public square, known as a "Viviero" or collective greenhouse, fostering community interaction. The design includes a gathering pavilion and a smaller terrace structure, constructed in five days using modular frames made from local Cryptomeria japonica with the help of community members.

Treasure Hill, an illegal high-density settlement of urban farmers in Taipei, in 2002 amidst demolition efforts by the city government. Notably, it demonstrated sustainable practices the city aimed to promote, such as greywater filtration, passive solar housing, and composting, while minimizing electricity use and car dependency. "Casagrande" discovered the place and halted its demolition, opting to restore the community's agricultural practices. Subsequently, Treasure Hill was recognized as a public environmental artwork, serving as a laboratory for sustainable urban living in Taipei.

Figure 2.20 depicts Treasure Hill Artist Village, a historic neighborhood in Taipei, Taiwan, initially developed as an illegal settlement by KMT veterans. It has been restored and preserved to exemplify sustainable urban living and artistic expression, characterized by its high-density, organic structure and "parasite urbanism," where new developments coexist with existing ruins. Today, it serves as a vibrant community that integrates residential homes, artist studios, and public spaces.

Fig. 2.20 Treasure Hill in Taipei (2002). *Source*: https://www.casagrandelaboratory.com

Cicada in Taipei serves as an organic refuge within the urban landscape, facilitating post-industrial transformation. Inspired by local knowledge and flexible bamboo structures, it embraces improvisation and an "insect mind." As a public space, it creates an oasis that isolates visitors from the surrounding city, allowing for introspection and connection with nature. Designed as an urban acupuncture project in 2011, Cicada aims to bridge modernity with ancient wisdom, symbolizing a harmonious relationship between people and their environment.

Figure 2.21 shows "Cicada," a bamboo pavilion by architect "Marco Casagrande" in Taipei, Taiwan. This cocoon-like structure, measuring 34 m long, 12 m wide, and 8 m high, functions as "urban acupuncture" to reconnect urban environments with nature. Built with traditional bamboo construction methods from Taiwan and South China, it serves as a public space for workshops, performances, and activities.

To synthesize these diverse examples, a broader pattern begins to emerge. Across different urban and cultural contexts, small, targeted spatial actions repeatedly demonstrate an ability to trigger disproportionate social and economic effects. What initially appear as isolated or experimental interventions collectively point toward a shared logic: when design strategically engages everyday practices and local conditions, it can recalibrate urban life without large-scale disruption. Empirical evidence for the efficacy of micro-scale interventions is found across multiple contexts. Tactical interventions have increased pedestrian activity, catalyzed local enterprise, and temporarily reduced vacancy rates (Lydon & Garcia, 2015). Larger but related micro-to-macro regeneration projects, such as the Cheonggyecheon stream restoration in Seoul, show how strategic removal and re-activation of urban fabric can transform mobility, environment, and civic life (Solecki & Rosenzweig, 2012). In

Fig. 2.21 Cicada in Taipei (2011). *Source*: https://www.casagrandelaboratory.com

Latin America, Medellín's social-urbanism investments (schools, cable cars, public spaces) illustrate how targeted public works combined with social programs can foster inclusion and reduce crime, though the scale and political contexts differ from acupuncture's modest interventions (Dávila, 2013).

Importantly, evidence also stresses limitations: without parallel measures addressing affordability and governance, such interventions can be appropriated by market interests and contribute to displacement or heritage commodification (Lees et al., 2008; Zukin, 2010). Therefore, when micro-design is integrated into inclusive institutional frameworks and connected to housing/affordability safeguards, the impact is strongest.

Evidence is primarily qualitative and case-based: documented revivals of underused courtyards, pedestrian flows increased by tactical insertions, and renewed stewardship after community co-design projects. Cite comparative evaluations that show short-term activation often precedes longer-term grassroots initiatives (project reports, post-occupancy observation).

2.4.4 Methodological and Ethical Considerations for Heritage Contexts

In heritage contexts, small interventions (restoring a historic bench, creating a mini-plaza, or installing informational signage) can increase public use and awareness of cultural assets with minimal cost. The literature describes urban acupuncture as a theory-based, bottom-up tactic for local improvement: it leverages community creativity and existing urban elements to catalyze broader regeneration without the need for large-scale projects (Casagrande, 2010).

Applying urban acupuncture in historic districts requires careful balance between innovation and conservation. Key considerations include:

- Cultural authenticity and sensitivity: Interventions must respect the site's heritage character to avoid commodification of intangible heritage. Designs should use compatible materials, colors and forms so as not to jar with the historic fabric. In practice, this means studying local building traditions and patterns ("genius loci") and ensuring new elements highlight, rather than overshadow, existing monuments (UNESCO, 2011). For example, seating or signage should reflect local aesthetics, and any artistic components might draw on traditional motifs. Stakeholders (heritage experts, local elders, shopkeepers) should be involved early so that projects reinforce shared identity and values (in line with heritage charters emphasizing socio-cultural authenticity).
- Community participation and co-design: Inclusive planning is essential. Urban acupuncture projects work best when residents help identify needs and co-create solutions. Participatory methods (workshops, charrettes, and surveys) can ensure that interventions address genuine local aspirations. Engaging artisans and youth groups, for instance, can yield context-aware designs and build stewardship. In heritage contexts, this also helps align with conservation goals: for example, locals may point out hidden historic features or propose uses (markets, festivals) that keep spaces alive. As "Casagrande notes", such projects are "more immediate and sensitive to community needs" than top-down schemes.
- Scale and reversibility: By definition, urban acupuncture favors small, reversible interventions. In heritage areas, this is an asset: minor changes can be tested without harming the site. Good practice is to pilot installations (temporary pavilions, pop-up gardens, and lighting trials) to gauge impact before permanent commitment. This flexibility also mitigates risk: if a design misfires, it can be undone with minimal cost. Moreover, small interventions often require less heavy machinery or demolition, reducing physical impact on fragile structures. The small-scale and often temporary nature of the intervention enables or even demands more careful construction, so that the surrounding historic fabric remains intact (Cutter et al., 2010).
- Avoiding disruption and gentrification: While aiming for revitalization, planners must avoid undermining heritage value or displacing vulnerable residents. Interventions should enhance usage (for example by improving lighting and seating in a square) without causing a dramatic property-value spike. In fact, because urban acupuncture is so localized, it tends to carry less risk of gentrification than large redevelopment (Balicka et al., 2021). Nonetheless, projects should be paired with measures like affordable housing support and local enterprise promotion to ensure benefits accrue to existing communities. Heritage sites often have strict

regulations; all projects must comply with preservation guidelines to prevent inadvertent damage (e.g. avoid anchoring anything into ancient walls) (UNESCO HUL approach).

- Integration into broader plans: Even micro-interventions should link to an overall heritage regeneration strategy. This means aligning projects with long-term goals (e.g. pedestrianizing an old market street or restoring a courtyard). Decision-support tools (such as scoring frameworks that emphasize cultural sustainability can help choose interventions that balance technical feasibility with socio-cultural priorities. Monitoring and documenting each pilot's outcomes is important: evidence from one intervention (say, adding benches along a historic promenade) can inform the next, creating a responsive planning cycle (Ghandehariun & Faghihi, 2025).
- Criteria for success: Good urban acupuncture in heritage contexts is place-based, incremental, and reversible. Projects should celebrate heritage assets (through interpretive signs, events or lighting) and fill genuine gaps in urban life (like a quiet square needing seating, or a blocked vista needing a new pedestrian link). They must be designed for community use—for example, installing shade trees and market stalls in a neglected plaza to draw families back. A useful criterion is whether an intervention is minimally invasive yet can "touch upon" many users' lives; another is whether it can be undone without trace if needed. By following these practices, urban acupuncture can refresh historic neighborhoods gently, keeping their unique character while addressing contemporary needs (Balicka et al., 2021).

It is important to emphasize that urban acupuncture is both a technical and ethical practice in heritage situations. Ethical design requires reflexivity about who benefits and who may be harmed by enhancement (Sandercock & Lyssiotis, 2003). Protection of intangible heritage requires protocols for consent and benefit sharing when local cultural practices or crafts are activated for visitors or commercial use (UNESCO, 2003).

Urban acupuncture should be intentionally designed to rebuild emotional (affect), cognitive (meaning/identity) and behavioral (use, care) bonds such as restoring social nodes, reintroducing rituals, or conserving landmarks that anchor memory. Use design cues (materials, scale, and program) that resonate with local narratives to re-establish belonging (Lewicka, 2011; Scannell & Gifford, 2010).

Ethical safeguards must assess risk of commodification/tourismification and include safeguards against accelerating displacement, anticipate and mitigate harm (e.g., involuntary exposure that increases tourist pressure) by pairing interventions with policy measures (affordable housing, local hiring) (UNESCO, 2011).Thus, acupuncture projects must align with local heritage policy, housing affordability measures, and governance reforms to avoid merely beautifying spaces for wealthier newcomers.

The parameters of analysis for urban acupuncture that will be utilized in the study and findings are displayed in Table 2.4.

Table 2.4 Set of parameters of urban acupuncture

Set of parameters of urban acupuncture		
Multi-functionality	Flexibility	Predictability

2.5 Concluding Summary of the Theoretical Framework

Urban acupuncture is a strategic, community-led micro-intervention that translates place-sensitive design principles into tactical acts that can rebuild attachment and catalyze social resilience in heritage contexts.

This chapter synthesized the interdisciplinary literature that frames gentrification's multifaceted pressures in historic and heritage contexts. The evolution of resilience from ecological persistence to social adaptability and transformability; the tripartite dimensions of social resilience—cohesion, adaptability, and recovery—and their operational relevance for heritage communities.

The psychological theory and measurement of place attachment as an emotional—cognitive—behavioral mediator linking design interventions to community outcomes; and the promise and constraints of urban acupuncture and tactical urbanism as situated, low-cost interventions that can reactivate public life without wholesale redevelopment.

From this review, three research implications emerge clearly. First, heritage contexts require a dual lens that treats tangible fabric and living culture as mutually constitutive: conservation that ignores everyday practices risks "museumification" and symbolic displacement. Second, place attachment is not a peripheral outcome but a central mechanism—measurable, actionable, and mediating—through which micro-scale design actions can translate into sustained resilience. Third, robust evaluation demands qualitative methods and accurate measurement to provide the coverage needed to capture material, institutional and psychosocial change.

Therefore, to synthesize the theoretical relationships discussed in this chapter, Fig. 2.22 presents the conceptual framework guiding this research. The model positions gentrification as a contextual pressure that disrupts residents' emotional, cognitive, and behavioral bonds to place, thereby weakening social resilience in heritage neighborhoods. Place attachment is conceptualized as a mediating psychological mechanism linking spatial change to collective resilience outcomes. Urban acupuncture is introduced as a targeted design intervention that operates through the restoration of place attachment, contributing to enhanced coping, adaptive, and transformative capacities within the community.

Figure 2.22 illustrates the conceptual apparatus of the study mapping the interrelations between gentrification pressures, urban acupuncture interventions, place attachment as a psychological mediator, and social resilience as a systemic outcome in heritage contexts.

This framework informs the analytical structure of the case studies presented in the next chapter.

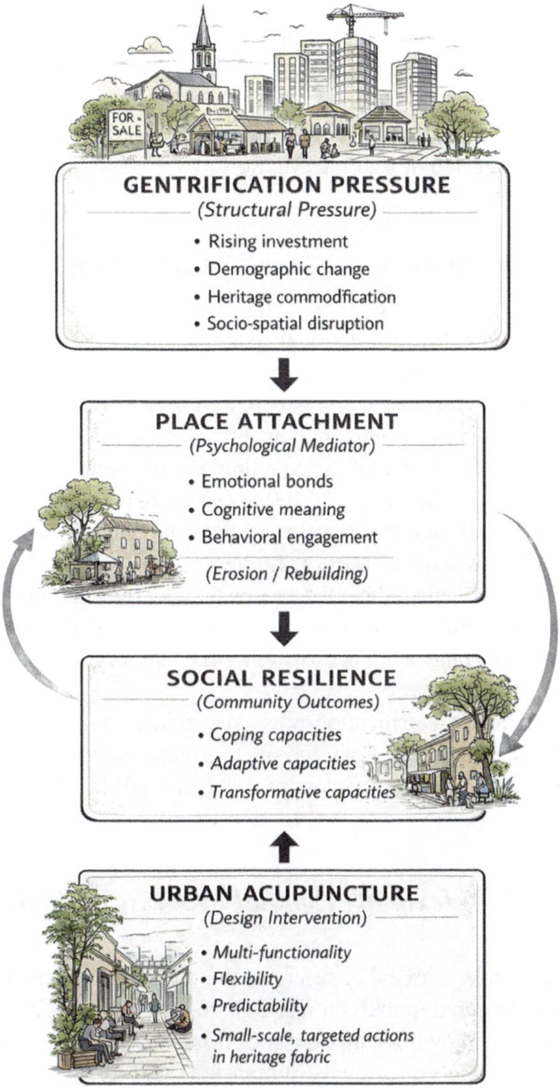

Fig. 2.22 The conceptual apparatus of the study. *Source*: Authors

2.6 Framework of Analysis

The framework of analysis translates the conceptual apparatus of the study (gentrification → place attachment → social resilience ← urban acupuncture) into six operational parameters used to examine and compare each case. Each parameter is described below through three complementary lenses:

(a) Theoretical rationale and positioning within the study's core concepts

(b) Pragmatic indicators suitable for contexts where direct surveying is not feasible (including observational, documentary, and photographic proxies)
(c) An explicit clarification of how the parameter mediates, moderates, or activates the relationships between gentrification pressures, place attachment dynamics, and social resilience outcomes, including the role of urban acupuncture as a targeted design intervention.

2.6.1 Multi-Functionality (Spaces that Serve Multiple Needs)

Rationale: Multi-functional spaces sustain everyday life under changing conditions and increase opportunities for diverse users to enact place-based practices. In heritage contexts, functional diversity helps preserve everyday cultural uses (markets, rituals, social meeting points) that anchor place attachment and reduce the purely commodified (tourist) use of space—thus contributing to by maintaining livelihoods and social ties (Norris et al., 2008; Zukin, 2010). Urban acupuncture interventions that add small facilities (seating, stalls, and shading) often multiply possible uses and therefore magnify impact (Lerner, 2014).

Observable indicators: Presence of mixed uses within a single space (e.g., trading, seating, cultural display), number of distinct programmed activities per week, photographic evidence of temporal variation in use, business mix data (local vs tourist).

Links: Gentrification pressures reduce multi-functionality when spaces convert to single-purpose high-end uses; urban acupuncture can restore multi-functionality, reinforcing place attachment (Lewicka, 2011).

2.6.2 Flexibility (Spatial Processes Adaptable to Change)

Rationale: Flexibility refers to the capacity of spatial systems and processes to be reconfigured quickly in response to new needs—e.g., converting a vacant lot into a market, allowing temporary performances in a courtyard. Flexibility supports adaptability (a core resilience dimension) and allows residents to continue meaningful practices despite socio-spatial change, preserving behavioral and cognitive aspects of place attachment (Magis, 2010; Walker et al., 2004). Urban acupuncture deliberately prioritizes reversible, low-cost solutions that enable flexibility in use and management (Lerner, 2014).

Observable indicators: Evidence of temporary/seasonal installations, availability of modular furniture/fixtures, municipal permissions for pop-ups, before/after photos showing reprogramming of space, records of event permits.

Links: In gentrifying areas, inflexible redevelopment accelerates displacement; flexible, small interventions can act as stopgaps that preserve use and attachment.

2.6.3 *Interactive Capacity (How Facilities Enable Resident–Environment Interactions)*

Rationale: Interactive capacity captures the degree to which design and amenities invite reciprocal engagement—social encounters, stewardship, sensory experiences—that constitute behavioral attachment and social capital (Perkins et al., 1996; Scannell & Gifford, 2010). Facilities that enable interaction (communal benches, markets, craft stalls) support everyday reciprocity and mutual aid, key elements of cohesion and coping capacity in resilience frameworks (Norris et al., 2008).

Observable indicators: Counts of social interactions (observed Clustering, number of informal gatherings), presence of community noticeboards or co-managed features, documented volunteer or stewardship initiatives, photographic sequences showing routine uses.

Links: Gentrification often reduces interaction by altering who uses a place; acupuncture measures aim to re-open channels for local interaction and everyday rituals.

2.6.4 *Multi-Locational Functionality (Facilities Able to Perform Similar Functions Across Sites)*

Rationale: Redundancy or multi-locational capacity (many small facilities that can substitute for each other) increase resilience by preventing single-point failures, if one node is lost or converted, others sustain function (Cutter et al., 2010). In heritage contexts, distributed small interventions (multiple pocket parks, several micro-markets) maintain opportunities for social life across the neighborhood and reduce the risk that one redevelopment will erase all communal functions. Redundancy also supports place attachment by preserving multiple loci of memory and practice.

Observable indicators: Mapping of similar facilities across the neighborhood (e.g., number of small parks, community kiosks), distance analysis (average spacing), photographic inventory of comparable public amenities, municipal service maps.

Links: Concentrated upscale redevelopment reduces redundancy; acupuncture strategies favor distributed, replicated micro-actions.

2.6.5 *Intelligent Public Services (Services that Respond to Complex, Changing Demands)*

Rationale: Intelligent services means service systems designed to be responsive and context-aware—integrating flexible delivery modes, local feedback, and contingency options. Such services (transport options, waste collection, markets scheduling) help communities cope with shocks and everyday stressors and are central to adaptive capacity (Folke, 2006). In heritage neighborhoods, services that recognize temporal rhythms (market days, religious processions) preserve cultural routines and thereby maintain place attachment. Urban acupuncture can act as a testing ground for service innovations (e.g., pilot waste-collection points, mobile market infrastructure).

Observable indicators: documented presence of adaptive service practices (mobile services, variable schedules), municipal pilot programs cited in reports, evidence of co-management agreements (NGO-municipality), service-usage logs if available.

Links: rigid, centralized services often fail heritage contexts; intelligent, small-scale service adaptations strengthen both attachment and resilience.

2.6.6 Responsiveness to Human Needs (Service Alternatives and Delivery Modes)

Rationale: This parameter emphasizes human-centeredness: the availability of alternatives in how services are provided and how public space supports diverse needs (accessibility, affordability, culturally appropriate programming). It is both an ethical criterion and a resilience asset: communities that can choose or adapt services retain dignity and agency, which supports psychosocial well-being and place attachment (Adger, 2000; Magis, 2010). Urban acupuncture projects that are co-designed with residents tend to be more responsive to local cultural requirements and therefore less likely to generate antagonistic gentrification outcomes.

Observable indicators: evidence of multiple service delivery options (e.g., formal market and informal vendors), records of participatory planning sessions, and photographic documentation of facilities adapted for different user groups (children, elders), fee structures showing affordable options.

Links: Lack of responsiveness is experienced as exclusion (weakening attachment); responsive design supports belonging and adaptive community actions.

The Table 2.5 shows the framework of analysis for this research:

This research adopts an analytical framework that translates theoretical constructs—gentrification, place attachment, social resilience, and urban acupuncture—into operational parameters for case study evaluation. Following the literature review, the framework synthesizes spatial, social, and psychological dimensions into a coherent set of analytical criteria. These parameters allow for a structured comparison of heritage-based urban acupuncture interventions and their capacity

Table 2.5 Framework of analysis for this research

Concepts	Set of parameters		
Social resilience	Coping capacities	Adaptive capacities	Transformative capacities
Urban acupuncture	Multi-functionality	Flexibility	Predictability
Place attachment	Cognitive	Emotional	Behavioral

to mitigate gentrification impacts, reinforce place attachment, and enhance social resilience. Rather than treating resilience and attachment as abstract outcomes, the framework evaluates how specific spatial and service-based characteristics support coping, adaptive, and transformative capacities within historic urban contexts.

Given the absence of direct user surveys or interviews, place attachment is assessed through spatial and behavioral proxy indicators, including continuity of use, community appropriation, frequency of recurrent public activities, symbolic visibility, patterns of repeated occupancy or stewardship and program persistence, consistent with established approaches in environmental psychology and heritage urban studies. These proxy indicators are observable from archival, spatial, and documentary sources.

This proxy approach is defensible in heritage research where material and behavioral traces reliably reflect cognitive and emotional bonds (Lewicka, 2011; Manzo & Perkins, 2006), but it should be reported as a limitation: proxies capture expressed and enacted attachment rather than self-reported affect, so findings about psychological intensity must be interpreted cautiously.

Table 2.5 analyzes therefore examine, in sequence, the social-resilience parameters, the urban acupuncture parameters, and the corresponding proxy measures of place attachment.

Figure 2.23 displays an analytical framework that links gentrification pressures to place attachment bonds and social resilience outcomes in heritage neighborhoods through urban design interventions. Gentrification pressures, influenced by economic and demographic changes, leads to socio-spatial disruption. Urban Acupuncture acts as a response strategy with small-scale, adaptable and multifunctional interventions. Place Attachment acts as a psychological mediator, linking spatial interventions to social outcomes through emotional belonging, cognitive meaning, and behavioral stewardship. Social resilience represents the community's ability to adapt to and transform in response to these changes.

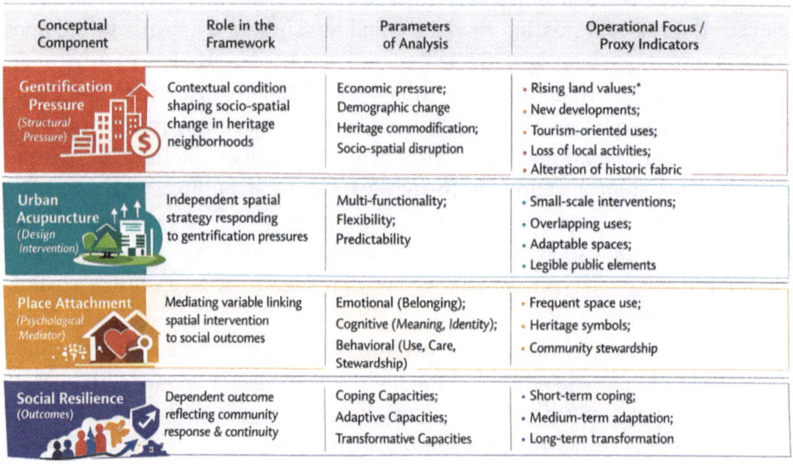

Analytical Framework Linking Gentrification, Place Attachment, Social Resilience, and Urban Acupuncture

Conceptual Component	Role in the Framework	Parameters of Analysis	Operational Focus / Proxy Indicators
Gentrification Pressure (Structural Pressure)	Contextual condition shaping socio-spatial change in heritage neighborhoods	Economic pressure; Demographic change Heritage commodification; Socio-spatial disruption	• Rising land values;* • New developments; • Tourism-oriented uses; • Loss of local activities; • Alteration of historic fabric
Urban Acupuncture (Design Intervention)	Independent spatial strategy responding to gentrification pressures	Multi-functionality; Flexibility; Predictability	• Small-scale interventions; • Overlapping uses; • Adaptable spaces; • Legible public elements
Place Attachment (Psychological Mediator)	Mediating variable linking spatial intervention to social outcomes	Emotional (Belonging); Cognitive (Meaning, Identity); Behavioral (Use, Care, Stewardship)	• Frequent space use; • Heritage symbols; • Community stewardship
Social Resilience (Outcomes)	Dependent outcome reflecting community response & continuity	Coping Capacities; Adaptive Capacities; Transformative Capacities	• Short-term coping; • Medium-term adaptation; • Long-term transformation

Fig. 2.23 The analytical framework linking gentrification, Place Attachment, Social Resilience and Urban Acupuncture for this research. *Source*: Authors

References

Aalbers, M. (2016). *Financialization of housing* (pp. 40–63). Oxfordshire, UK: Taylor & Francis.

Adger, W. N. (2000). Social and ecological resilience: Are they related? *Progress in Human Geography, 24*(3), 347–364.

Ahern, J. (2011). From fail-safe to safe-to-fail: Sustainability and resilience in the new urban world. *Landscape and Urban Planning, 100*(4), 341–343.

Al-Hinkawi, W. S., & Al-Saadi, S. M. (2020). Urban acupuncture, a strategy for development: case study of al-rusafa, Baghdad. In *IOP conference series: Materials science and engineering* (Vol. 881, No. 1, p. 012002). IOP Publishing.

Al-Mendilawi, M. M. A., & Al-Saaidy, H. J. E. (2024). Effect of urban spaces in enhancing place attachment in vertical residential complexes (Bismayah as a case study). *Civil Engineering and Architecture, 12*(4), 2519–2531.

Altman, I., & Low, S. (1992). *Place attachment: Human behavior and environment*. Plenum Press.

Andersson, Å. E., & Andersson, D. E. (2017). *Time, space and capital*. Edward Elgar Publishing.

Balicka, J., Storie, J. T., Kuhlmann, F., Wilczyńska, A., & Bell, S. (2021). Tactical urbanism, urban acupuncture and small-scale projects. In *Urban blue spaces* (pp. 406–430). Routledge.

Beatley, T. (1998). The vision of sustainable communities. *Cooperating with nature: Confronting natural hazards with land-use planning for sustainable communities* (pp. 233–262).

Béné, C., Wood, R. G., Newsham, A., & Davies, M. (2012). Resilience: new utopia or new tyranny? Reflection about the potentials and limits of the concept of resilience in relation to vulnerability reduction programmes. *IDS Working Papers, 2012*(405), 1–61.

Berkes, F., Colding, J., & Folke, C. (2003). *Navigating social-ecological systems. Building resilience for complexity and change*. Cambridge.

Bonaiuto, M., Fornara, F., & Bonnes, M. (2003). Indexes of perceived residential environment quality and neighbourhood attachment in urban environments: A confirmation study on the city of Rome. *Landscape and Urban Planning, 65*(1–2), 41–52.

Brandt, J., & Vejre, H. (2004). Multifunctional landscapes—advances in ecological planning. *Landscape and Urban Planning*.

Carpenter, S., Walker, B., Anderies, J. M., & Abel, N. (2001). From metaphor to measurement: Resilience of what to what? *Ecosystems, 4*(8), 765–781.
Casagrande, M. (2010). Urban acupuncture. *Retrieved February 29*, 2012.
Casagrande, M. (2019). From urban acupuncture to the third generation city. *Nature driven urbanism* (pp. 131–153). Springer International Publishing.
Casprini, D., Oppio, A., Di Tommaso, A., Datola, G., Dell'Ovo, M., Torrieri, F., & Rossitti, M. (2026). The impact of urban acupuncture: Adopting a social innovation lens to assess the value of localized urban intervention. *Cities, 169*, 106502.
Colombia's Medellin rides cable car to a better future. 27 January 2014. Retrieved from https://www.taipeitimes.com/News/world/archives/2014/01/27/2003582279
Cutter, S. L., Burton, C. G., & Emrich, C. T. (2010). The landscape of disaster resilience indicators in the USA. *Natural Hazards, 55*(1), 1–22. https://doi.org/10.1007/s11069-009-9410-2
Davidson, M. (2009). Displacement, space and dwelling: Placing gentrification debate. *Ethics Place and Environment (Ethics, Place & Environment (Merged with Philosophy and Geography)), 12*(2), 219–234.
Dávila, A. L. (2013). *La economía del conocimiento y el socialismo*. Sello Editorial Academia.
Desmond, M. (2017). *Evicted: Poverty and profit in the American city*. Crown.
Dikmak, A., Farahat, B., & Mohsen, H. (2024). Urban acupuncture as recovery approach to improving the liveability of neighbourhoods affected by Beirut Port explosion. *Rigas Tehniskas Universitates Zinatniskie Raksti, 20*(1), 124–139. https://doi.org/10.13189/cea.2024.120402
Fabbricatti, K., Boissenin, L., & Citoni, M. (2020). Heritage community resilience: Towards new approaches for urban resilience and sustainability. *City, Territory and Architecture, 7*(1), 17.
Fleming, J., & Ledogar, R. J. (2008). Resilience, an evolving concept: A review of literature relevant to Aboriginal research. *Pimatisiwin, 6*(2), 7.
Florida, R. (2002). *The rise of the creative class: And how it's transforming work, leisure, community and everyday life*. Basic Books.
Folke, C. (2006). Resilience: The emergence of a perspective for social–ecological systems analyses. *Global Environmental Change, 16*(3), 253–267. https://doi.org/10.1016/j.gloenvcha.2006.04.002
Folke, C., Carpenter, S. R., Walker, B., Scheffer, M., Chapin, T., & Rockström, J. (2010). Resilience thinking: integrating resilience, adaptability and transformability. *Ecology and Society, 15*(4).
Foth, M. (Ed.). (2008). *Handbook of research on urban informatics: The practice and promise of the real-time city: the practice and promise of the real-time city*. IGI Global.
Gehl, J. (2010). *Cities for people*. Island Press.
Ghandehariun, S., & Faghihi, S. V. (2025). Dynamic decision-support for heritage-sensitive urban regeneration: An AI-driven framework integrating socio-cultural sustainability. *Results in Engineering, 28*, 108189.
Gifford, R. (2007). *Environmental psychology: Principles and practice* (4th ed.). Optimal Books.
Glaeser, E. (2012). *Triumph of the city: How our greatest invention makes us richer, smarter, greener, healthier, and happier*. Penguin.
Godschalk, D. R. (2003). Urban hazard mitigation: Creating resilient cities. *Natural hazards review*, 4(3), 136–143.
Guerra, M. I. (2014). *Regulating neglect: Territory, planning, and social transformation in Medellín*. University of California, Berkeley.
Gunderson, L. H., & Holling, C. S. (2002). *Panarchy: Understanding transformations in human and natural systems* (pp. xxiv+-507).
Hackworth, J., & Smith, N. (2001). The changing state of gentrification. *Tijdschrift Voor Economische En Sociale Geografie, 92*(4), 464–477.
Hamdi, N. (2010). *The placemaker's guide to building community*. Routledge.
Hammoodi, S. A., & Al-Hinkawi, W. S. (2023). The role of spatial value in the reconstruction of religious buildings Mosul city—A case study. *Ain Shams Engineering Journal, 14*(10), 102164.
Holling, C. S. (1973). *Resilience and stability of ecological systems*.

Holling, C. S. (1996). Surprise for science, resilience for ecosystems, and incentives for people. *Ecological Applications, 6*(3), 733–735.

Holling, C. S. (2001). Understanding the complexity of economic, ecological, and social systems. *Ecosystems, 4*(5), 390–405.

Keck, M. (2014). Resilience potentials of traditional food retailers in Dhaka against the backdrop of emerging supermarkets. *In Resilience in the social sphere: Theoretical and empirical analyses* (pp. 181–206). Springer Fachmedien Wiesbaden.

Kourtit, K., Nijkamp, P., Türk, U., & Wahlstrom, M. (2022). City love and neighbourhood resilience in the urban fabric: A microcosmic urbanometric analysis of Rotterdam. *Journal of Urban Management, 11*(2), 226–236.

Kwok, A. H., Doyle, E. E., Becker, J., Johnston, D., & Paton, D. (2016). What is 'social resilience'? Perspectives of disaster researchers, emergency management practitioners, and policymakers in New Zealand. *International Journal of Disaster Risk Reduction, 19*, 197–211.

Kyle, G. T., Mowen, A. J., & Tarrant, M. (2004). Linking place preferences with place meaning: An examination of the relationship between place motivation and place attachment. *Journal of Environmental Psychology, 24*(4), 439–454.

Kyle, G., Graefe, A., & Manning, R. (2005). Testing the dimensionality of place attachment in recreational settings. *Environment and Behavior, 37*(2), 153–177.

Lees, L., Slater, T., & Wyly, E. (2008). *Gentrification*. Routledge. https://doi.org/10.4324/9780203940877

Lerner, J. (2014). *Urban acupuncture: Celebrating pinpricks of change that enrich city life*. Lars Müller Publishers.

Lewicka, M. (2011). Place attachment: How far have we come in the last 40 years? *Journal of Environmental Psychology, 31*(3), 207–230. https://doi.org/10.1016/j.jenvp.2010.10.001

Lorenz, D. F. (2013). The diversity of resilience: Contributions from a social science perspective. *Natural Hazards, 67*(1), 7–24.

Lydon, M., & Garcia, A. (2015). A tactical urbanism how-to. In *Tactical urbanism: Short-term action for long-term change* (pp. 171–208). Island Press/Center for Resource Economics.

Maciag, M. (2015a). Millennials Remade Cities, But Will They Keep Living in Them? In *Governing the States and Localities*.

Maciag, M. (2015b). *Gentrification in America report*. Governing. Retrieved from https://www.governing.com/archive/gentrification-in-cities-governing-report.html

Magis, K. (2010). Community resilience: An indicator of social sustainability. *Society and Natural Resources, 23*(5), 401–416.

Manzo, L., & Devine-Wright, P. (2013). *Place attachment: Advances in theory, methods and applications*. Routledge.

Manzo, L. C., & Perkins, D. D. (2006). Finding common ground: The importance of place attachment to community participation and planning. *Journal of Planning Literature, 20*(4), 335–350. https://doi.org/10.1177/0885412205286160

Marcus, F. (2009). *Handbook* of *research on urban informatics: The practice and promise of the real-time city*. In *Information science reference* (pp. 17–19). IGI Global.

Masten, A. S. (2001). Ordinary magic: Resilience processes in development. *American Psychologist, 56*(3), 227.

Meerow, S., Newell, J. P., & Stults, M. (2016). Defining urban resilience: A review. *Landscape and Urban Planning, 147*, 38–49.

Morales D. (2004). The strategy of urban acupuncture: *Structure fabric and topography conference*. Nanjing University, China (pp. 55–56).

Norman, D. A. (2013). *The design of everyday things*. Basic Books.

Norris, F. H., Stevens, S. P., Pfefferbaum, B., Wyche, K. F., & Pfefferbaum, R. L. (2008). Community resilience as a metaphor, theory, set of capacities, and strategy for disaster readiness. *American Journal of Community Psychology, 41*, 127–150. https://doi.org/10.1007/s10464-007-9156-6

O'Brien, K., Hayward, B., & Berkes, F. (2009). Rethinking social contracts: building resilience in a changing climate. *Ecology and society, 14*(2).

Obrist, B., Pfeiffer, C., & Henley, R. (2010). Multi-layered social resilience: A new approach in mitigation research. *Progress in Development Studies, 10*(4), 283–293.

ORIZZONTALE - CASA DO QUARTEIRÃO. Retrieved from https://divisare.com/projects/332350-orizzontale-rui-soares-casa-do-quarteirao

Peláez, F. (2016). Colonia Roma, Mexico City (2016). CC-BY-SA 3.0, Wikimedia Commons / Mexico City's Roma neighbourhood gentrification. Retrieved from https://commons.wikimedia.org/wiki/File:Mexico_City%27s_Roma_neighborhood_gentrification_2016.jpg

Pelling, M., & Manuel-Navarrete, D. (2011). From resilience to transformation: the adaptive cycle in two Mexican urban centers. *Ecology and society, 16*(2).

Pendlebury, J. (2013). Conservation values, the authorised heritage discourse and the conservation-planning assemblage. *International Journal of Heritage Studies, 19*(7), 709–727.

Pennock, S. F. (2017). *Resilience in positive psychology: How to bounce back.* Retrieved from https://positivepsychology.com/resilience-in-positive-psychology/

Place Attachment: An insight system to create lasting communities. July 2021. Retrieved from https://mindfolio.com/place-attachment-insight-system-to-create-lasting-communities

Perkins, D. D., Brown, B. B., & Taylor, R. B. (1996). The ecology of empowerment: Predicting participation in community organizations. *Journal of Social Issues, 52*(1), 85–110.

Perkins, D. D., Hughey, J., & Speer, P. W. (2002). Community psychology perspectives on social capital theory and community development practice. *Community Development, 33*(1), 33–52.

Perrings, C. (1998). Resilience in the dynamics of economy-environment systems. *Environmental and Resource Economics, 11*(3), 503–520.

Proshansky, H. M., Fabian, A. K., & Kaminoff, R. (1983). Place-identity: Physical world socialization of the self. *Journal of Environmental Psychology, 3*(1), 57–83. https://doi.org/10.1016/S0272-4944(83)80021-8

Resilience. (2019). *Thrive Ottawa County.* Retrieved from https://thriveottawa.org/resilience-2/

Rosenthal Park I Lane Cove, Australia I Corkery Consulting. Retrieved from https://worldlandscapearchitect.com/rosenthal-park-lane-cove-australia-corkery-consulting/?v=26403ec6d537

Safeguarding our living heritage. UNESCO Intangible Cultural Heritage. Retrieved from https://ich.unesco.org

Sandercock, L., & Lyssiotis, P. (Eds.). (2003). *Cosmopolis II: Mongrel cities of the 21st century.* A&C Black.

Sankt Kjeld' Square & Bryggervangen. *A Copenhagen landmark for nature-based cloudburst protection combined with recreational spaces, biodiversity, and new infrastructure.* Retrieved from https://www.sla.dk/cases/sankt-kjelds-square-and-bryggervangen/

Scannell, L., & Gifford, R. (2010). Defining place attachment: A tripartite organizing framework. *Journal of Environmental Psychology, 30*(1), 1–10. https://doi.org/10.1016/j.jenvp.2009.09.006

Shaw, K. (2008). Gentrification: What it is, why it is, and what can be done about it. *Geography Compass, 2*(5), 1697–1728.

Sharifi, A., & Yamagata, Y. (2018). Resilience-oriented urban planning. In *Resilience-oriented urban planning: theoretical and empirical insights* (pp. 3–27). Cham: Springer International Publishing.

Smith, N. (1979). Toward a theory of gentrification a back to the city movement by capital, not people. *Journal of the American Planning Association, 45*(4), 538–548.

Smith, N. (2002). New globalism, new urbanism: Gentrification as global urban strategy. *Antipode, 34*(3), 427–450.

Smith, L. (2006). *Uses of heritage.* Routledge.

Solecki, W. D., & Rosenzweig, C. (2012). *Cheonggyecheon Stream Restoration Case Study.* https://www.archdaily.com/1020945/re-naturalization-of-urban-waterways-the-case-study-of-cheonggye-streamin-seoul-south-korea

Steinberg, M., & Burby, R. J. (2002). Growing safe. *Planning, 68*(4), 22–22.

Stokols, D., & Shumaker, S. A. (1982). The psychological context of residential mobility and Weil-Being. *Journal of Social Issues, 38*(3), 149–171.

Thurber, A., & Krings, A. (2021). *Gentrification*. Oxford University Press. https://doi.org/10.1093/acrefore/9780199975839.013.1413
Till, J., & Schneider, T. (2005). Flexible housing: the means to the end. ARQ: *architectural research quarterly, 9*(3–4), 287–296.
UNESCO. (2003). *Convention for the safeguarding of the intangible cultural heritage*. UNESCO. https://ich.unesco.org/en/convention
UNESCO. (2011). *Recommendation on the historic urban landscape*. United Nations Educational, Scientific and Cultural Organization. https://whc.unesco.org/en/hul/
Urban Bloom / AIM Architecture + URBAN MATTERS. Retrieved from https://www.archdaily.cl/cl/893837/urban-bloom-aim-architecture-plus-urban-matters
Urban Acupuncture. October 8, 2024. Retrieved from https://alchetron.com/
Vale, L. J. (2014). The politics of resilient cities. *Building Research & Information, 42*(2), 191–201.
Vale, L. J., & Campanella, T. J. (Eds.). (2005). *The resilient city: How modern cities recover from disaster*. Oxford University Press.
Voss, M. (2008). The vulnerable can't speak. An integrative vulnerability approach to disaster and climate change research. *Behemoth—A Journal on Civilisation, 1*(3), 39–56.
Waights, S. (2019). The preservation of historic districts—Is it worth it? *Journal of Economic Geography, 19*(2), 433–464.
Walker, B., Carpenter, S., Anderies, J., Abel, N., Cumming, G., Janssen, M., ... & Pritchard, R. (2002). Resilience management in social-ecological systems: a working hypothesis for a participatory approach. *Conservation ecology, 6*(1).
Walker, B., Holling, C. S., Carpenter, S. R., & Kinzig, A. (2004). Resilience, adaptability and transformability in social–ecological systems. *Ecology and Society, 9*(2), 5.
Walker, B., & Salt, D. (2012). *Resilience thinking: sustaining ecosystems and people in a changing world*. Island press.
West, D. (2011). *Urban acupuncture' touted for cash-strapped cities*. New Urban Network.
Williams, D. R., & Vaske, J. J. (2003). The measurement of place attachment: Validity and generalizability of a psychometric approach. *Forest science, 49*(6), 830–840.
Zukin, S. (2009). *Naked city: The death and life of authentic urban places*. Oxford University Press.
Zukin, S. (2010). *Naked city: The death and life of authentic urban places*. Oxford University Press.

Chapter 3
Case Studies: Applying Urban Acupuncture as an Architectural Strategy

This chapter examines the practical application of urban acupuncture as a design-led strategy for strengthening place attachment and enhancing social resilience within heritage contexts affected by gentrification. Through three carefully selected case studies—representing diverse socio-spatial contexts in which small-scale, targeted architectural interventions have generated broader urban, social, and psychological impacts—the chapter translates the theoretical framework developed in earlier chapters into empirical analysis, allowing abstract concepts to be tested against real spatial, social, and cultural conditions.

Each case study is analyzed using a consistent qualitative methodology based on the study's core parameters: urban acupuncture (multi-functionality, flexibility, predictability), place attachment (emotional, cognitive, and behavioral dimensions), and social resilience (coping, adaptive, and transformative capacities). This structured approach enables systematic comparison while remaining sensitive to contextual differences in scale, governance, heritage value, and socio-economic conditions.

By situating each intervention within its broader gentrification dynamics, the chapter reveals how small-scale, targeted architectural and urban actions can mediate psychosocial disruption, re-anchor community identity, and support resilient patterns of everyday life. Collectively, the case studies provide critical insight into the conditions under which urban acupuncture operates most effectively, setting the foundation for the comparative analysis, discussion, and design-oriented recommendations that follow.

The findings contribute to a deeper understanding of how architectural precision, strategic placement, and human-centered design can function as effective tools for urban regeneration and socio-spatial healing.

B. I. Farahat and H. O. Alaeddine, *Urban Acupuncture for Place Attachment and Social Resilience in Heritage Contexts*, Creativity, Heritage and the City 10,
https://doi.org/10.1007/978-981-95-8970-8_3

3.1 Methods of Analyzing Case Studies

This study uses a qualitative, comparative case-study method to examine how urban acupuncture interventions interact with place attachment to produce social-resilience outcomes in heritage neighborhoods. The analytical procedure is structured and repeatable so that each case is evaluated against the same parameter set and proxy indicators described in the theoretical framework. The parameters and proxy indicators established in the theoretical section serve as the foundation for the analysis: Social resilience (coping capacities, adaptive capacities, transformative capacities); Urban Acupuncture (multi-functionality, flexibility and predictability); Place attachment (emotional, cognitive and behavioral).

- Analytical steps:

   a. Descriptive mapping: an ID card and spatial diagram are produced for the case (location, date, scale, intervention points, and program).
   b. Parameter coding: systematically code evidence against the three-parameter groups: Social Resilience (coping/adaptive/transformative), Urban Acupuncture (multi-functionality/flexibility/predictability), and Place Attachment (emotional/cognitive/behavioral). Codes use a shared codebook so assessments are comparable across cases.
   c. Proxy operationalization: where direct survey data are absent, infer psychosocial constructs via proxies (e.g., repeated use = behavioral attachment; presence of heritage markers = cognitive attachment; number of local events/vendor days = coping/adaptive proxies).
   d. Within-case assessment: a concise analytical narrative is written for each case that links intervention features to place attachment proxies and resilience outcomes, noting enabling or constraining contextual factors (governance, market pressure, cultural practices).
   e. Cross-case comparison: populated through the comparative matrix (see Table 3.6) to synthesize patterns of similarity and difference.

- Presentation of results:

   Each case includes an ID card, key figures (site plan/diagrams/photos), the parameter-by-parameter analysis (definition → evidence/proxy → assessment), and a short conclusion table summarizing the assessment and confidence levels.

3.2 Criteria of Selection

The case studies were chosen based on several criteria. The first criterion is how urban acupuncture is used differently, the second criterion is how the neighborhood is different, and the strategy's degree of extension serves as the basis for the third criterion.

Case study selection followed purposive, theory-driven sampling to maximize analytical contrast and ensure that each case meaningfully tests components of the conceptual framework. The aim was not representativeness but analytical leverage: selecting cases that illuminate different pathways through which urban acupuncture can affect place attachment and resilience.

3.3 Methodological-Conceptual Structure

To ensure analytical clarity and comparability across the selected case studies, each parameter is examined through a consistent three-step structure: (1) definition and theoretical rationale, (2) application to the specific case, and (3) assessment based on observable evidence or proxy indicators. This structured approach enables the concepts of this research (social resilience, urban acupuncture, and place attachment) to be systematically translated into spatial and operational terms that can be examined within urban contexts.

Given the constraints of direct surveying and longitudinal social data, the analysis relies on design-based, observational, and documentary proxies, including spatial configuration, programmatic diversity, patterns of use, temporal adaptability, and photographic documentation. Such proxy indicators are widely adopted in architectural and urban research as valid means of interpreting socio-psychological processes when direct user data are unavailable. By moving from theory to situated application and finally to evaluative judgment, this method ensures conceptual coherence, methodological transparency, and a robust comparative framework for assessing how small-scale urban interventions function within heritage environments under gentrification pressure.

This research contributes a design-led analytical framework that positions place attachment as a psychological mediator between gentrification pressures and social resilience outcomes, operationalized through urban acupuncture strategies in heritage contexts.

3.4 Analyzing Case Study I—LibreLigne—Urban Acupuncture in Libreville, Gabon (2014)

LibreLigne is a conceptual urban-acupuncture proposal that reconnects Libreville with its seafront by concentrating identity, programming and public life into five pier-like nodes (jetées) along the promenade. Rather than diffuse interventions along the whole coastline, the project concentrates visual impact and civic activity in specific nodes to catalyze broader shoreline activation, improve access (pedestrian and cycling), and support community programming.

Table 3.1 ID card for LibreLigne—Gabon

LibreLigne ID card	
See Fig. 3.1	
Name	LibreLigne
Location	Libreville, Gabon
Proposal date	May, 2014
Architect/s	Ecosistema Urbana
Typology	Pier-like public nodes (jetées); continuous waterfront promenade
Architecture trend	Pier-like structures (jetées)

Fig. 3.1 LibreLigne project overview, Gabon. Conceptual visualization of the LibreLigne urban acupuncture strategy, illustrating the concentration of public life and identity at discrete waterfront nodes (jetées) along the coastal promenade. *Source*: ecosistema urbana (2014)

The major purpose of the idea is to reconnect Libreville with the seaside and to extend the city's life to the waterfront in certain areas. This would be accomplished by urban regeneration operations tailored to the existing as well as the specific cultural, sociological, and economic setting. Instead of spreading them out throughout the promenade, the concept focuses visual impact, identity, and activity at five nodes (Table 3.1).

3.4.1 Analyzing Parameter I—Coping Capacities

Definition & rationale: Coping capacities are the immediate resources, spaces and practices that individuals and communities can mobilize to manage short-term stresses or disturbances (UNDRR, 2015). In urban contexts, these include places for mutual aid, short-term economic activity and information exchange.

Application to LibreLigne: The LibreLigne concept explicitly creates a set of discrete, legible public nodes (jetées) designed to host markets, cultural events and gatherings. By concentrating activity at identifiable piers, the proposal provides spatial platforms where citizens could gather quickly, share information, distribute aid or mount short-term economic responses (pop-up markets, informal vending) during episodic shocks. The promenade has proposed pedestrian and bike connections further support rapid mobilization across adjacent neighborhoods.

Evidence/proxies: As a proposal rather than a completed project, evidence is limited to design drawings and programmatic statements; useful proxies include documented event programming, photographic records of pilot activations (if any), and plans showing circulation links (Ecosistema Urbano, 2014). These proxies indicate moderate potential for coping support but require on-the-ground testing to confirm actual use under stress.

Assessment: Conceptually strong: the design foregrounds short-term social and economic functions that support coping. However, without implementation data (event logs, usage counts), claims remain speculative and should be tested with observational studies or pilot activations (UNDRR, 2015).

Figure 3.2 shows the proposed shared-space crossings in front of the jetées, designed to enhance accessibility, everyday coping capacity, and informal social interaction between pedestrians and motorists. It introduces a sustainable design called "urban prosthesis," integrating housing, offices, commercial, and public spaces. Key features include modern infrastructure like wind turbines for renewable energy and green structures.

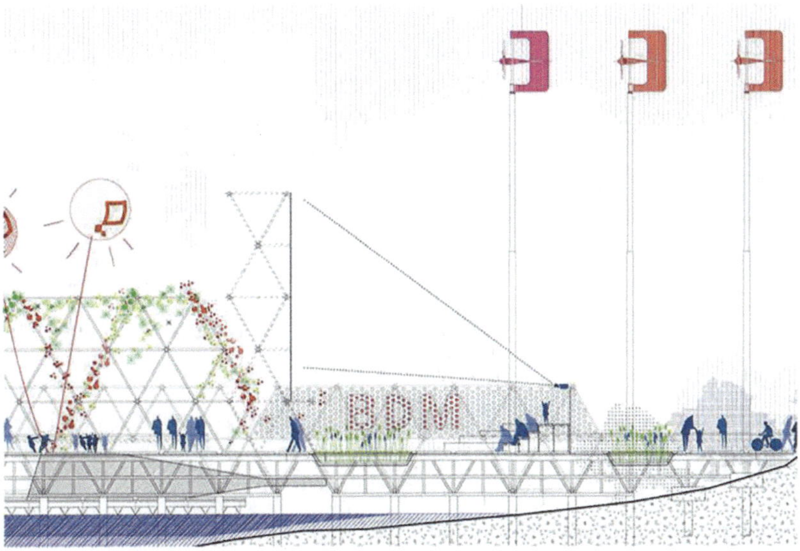

Fig. 3.2 The shared areas for pedestrian and vehicular crossings. *Source*: ecosistema urbana (2014)

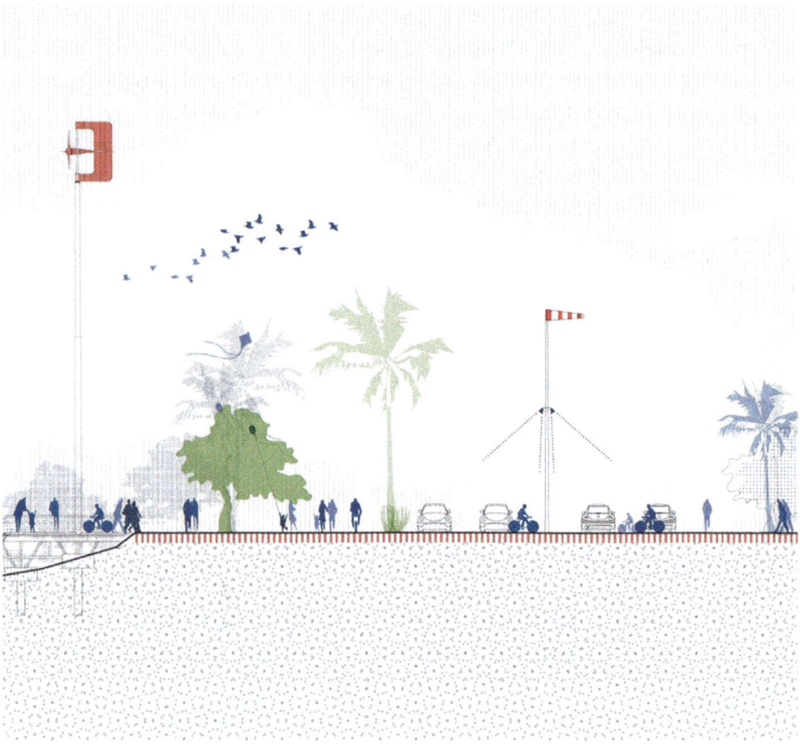

Fig. 3.3 Temporal transformation of uses. *Source*: ecosistema urbana (2014)

Figure 3.3 shows how activities on the jetées shift throughout the day, demonstrating adaptability and multi-functionality in response to changing social rhythms. Key features of the design encompass renewable energy through wind turbines, recreational spaces including pedestrian walkways and cycling paths, enhanced urban connectivity to the coast, and public amenities that promote social interaction and activity along the bay.

3.4.2 Analyzing Parameter II—Adaptive Capacities

Definition & rationale: Adaptive capacity refers to the ability of a place and its institutions to adjust practices, uses and management in response to new conditions (UNDP, 2003). It includes modularity, resource diversity and institutional learning.

Application to LibreLigne: The modular jetée typology and continuous promenade (with bike lanes and flexible program areas) provide physical conditions favorable to adaptation: nodes can host different functions over time and the linear system allows reconfiguration of activities along the shoreline. The thematic programming (culture, music, education, nature) offers a template that can be reallocated in response to shifting community needs.

Evidence/Proxies: Relevant proxies include the jetée design variations (sections showing different profiles), presence of multiuse spaces in diagrams, and the inclusion of a civic app to coordinate activities—these signal intentional flexibility. Analysis of these design artefacts shows substantive adaptive potential, particularly at the programmatic level (Till & Schneider, 2005).

Assessment: High adaptive potential on paper; the proposal's modularity and programming logic enable incremental adjustments. Real-world adaptive capacity will depend on governance arrangements and the ability to iterate programming with local stakeholders.

Figure 3.4 shows the plan view of the Libreville coastline highlighting the modular structure of the promenade and jetées, supporting adaptive and flexible spatial configurations. The plan outlines intervention points along the coastline to enhance social interaction, employing urban regeneration strategies suited to local cultural and economic contexts, and aims to integrate urban life with the seashore.

3.4.3 Analyzing Parameter III—Transformative Capacities

Definition & rationale: Transformative capacity involves deeper changes in institutions, governance, and social values that realign systems and open new developmental trajectories (Walker et al., 2004).

Application to LibreLigne: While primarily spatial, LibreLigne includes civic engagement mechanisms (public programming, a mobile app) that could catalyze new stewardship practices and place-based governance for the waterfront. If coupled with policy instruments (affordable vendor spaces, community management frameworks), such spatial projects can trigger institutional shifts in waterfront management and local economic organization.

Evidence/Proxies: Proxies include explicit references in project materials to civic platforms and programmed governance activities; however, there is no evidence of formal policy adoption or institutional commitments. Longitudinal documents (policy briefs, municipal schedules) would be necessary to demonstrate transformation.

Assessment: Transformative capacity is plausible but conditional: spatial proposals create openings for institutional change, yet transformation requires complementary policy and capacity building—which the project proposal suggests but does not secure.

Figure 3.5 shows a sectional drawing of the largest pier, illustrating its capacity to host multiple functions and to act as a catalyst for waterfront transformation.

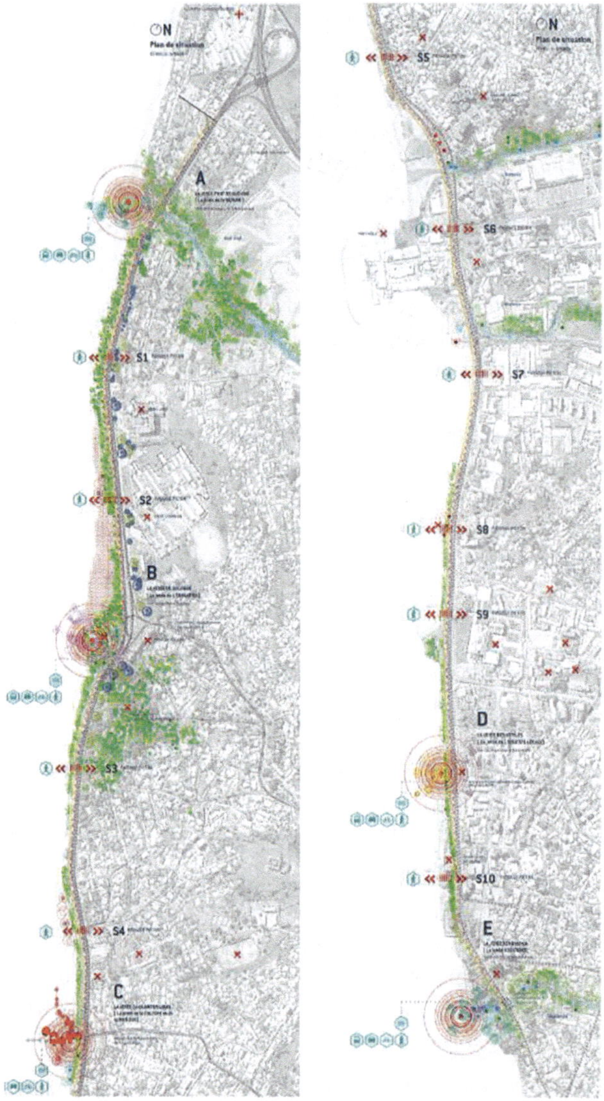

Fig. 3.4 Plan of the coastal promenade line of Libreville, in two sections. *Source*: ecosistema urbana (2014)

The design aims to reconnect the city with the seashore through urban regeneration, featuring unique structural landmarks that enhance citizen orientation and extend urban life to the waterfront.

The linear promenade, on the other hand, alters the layout of its segment based on the surroundings. In that part, some important terms are described, including

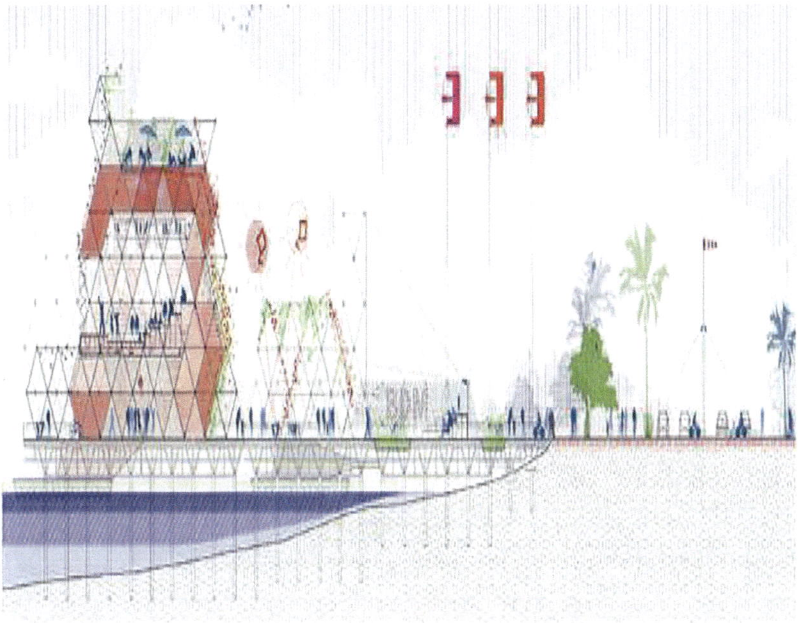

Fig. 3.5 Section of the main 'Jetée'. *Source*: ecosistema urbana (2014)

"waterfront avenue for (unavoidable) motorized traffic," "series of landscape markers" (related to energy production and visibility), "coastal bike lane," "waterfront promenade," and "urban appropriable fringe". Combining, stretching, or contracting these parts results in a variety of profiles and interest regions.

3.4.4 Analyzing Parameter IV—Multi-Functionality

Definition & rationale: Multi-functionality denotes the capacity of places to support diverse activities across times of day and user groups, thereby reducing vulnerability tied to single uses (Brandt & Vejre, 2004).

Application to LibreLigne: The five themed jetées are explicitly designed to host simultaneous and sequential uses (markets, concerts, education, and recreational activity). Sections and plans show seating, platforms and service zones that encourage overlapping activities. This design logic reduces single-use dependency and increases daily activation.

Evidence/Proxies: Plans and sectional drawings indicate provision for multiple activity types and service points. Photographic simulations (project renders)

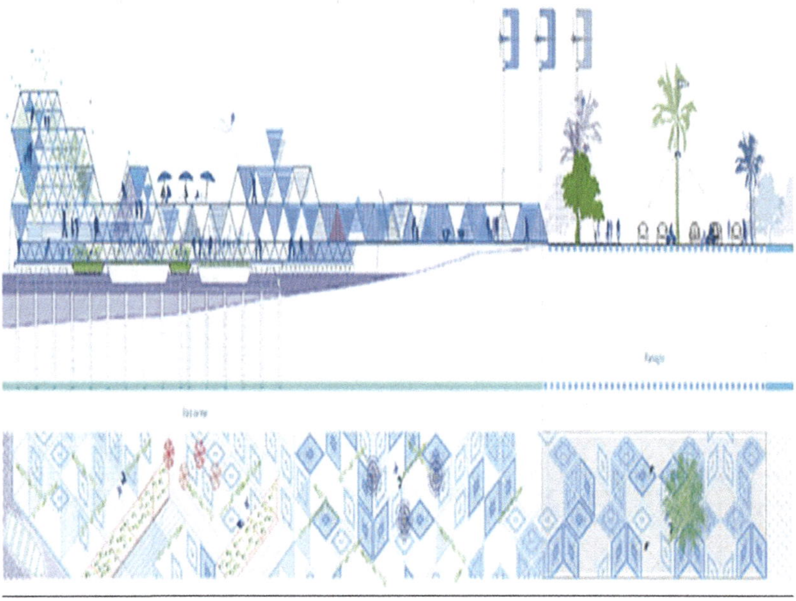

Fig. 3.6 Section of one of a lower pier, extended structure. *Source*: ecosistema urbana (2014)

show day-to-day and evening uses—these serve as reasonable proxies for multi-functionality potential.

Assessment: Strong design intent and high theoretical capacity for multi-functionality. Implementation would need to ensure infrastructure (shade, seating, utilities) and programming to realize this capacity.

Figure 3.6 shows a lower-profile jetée designed to accommodate diverse activities, reinforcing multi-functionality and user-driven appropriation. The design presents a modular framework that can expand or contract depending on contextual requirements, integrating didactic and informational spaces with public leisure areas. It is intended to be flexible for different settings, including beaches and urban abandoned lots, with the goal of enhancing visibility and regulation of waterfront activities.

3.4.5 *Analyzing Parameter V—Flexibility*

Definition & rationale: Flexibility refers to physical and programmatic configurability that allows space to be repurposed with low cost and minimal disruption (Till & Schneider, 2005).

Fig. 3.7 Section of reduced promenade along the beach area. *Source*: ecosistema urbana (2014)

Application to LibreLigne: The jetée typology, with varying lengths and profiles and the use of lightweight interventions, supports reconfiguration (temporary stalls, pop-ups, stage platforms). The promenade's ability to narrow or widen in plan allows circulation and activity zones to be shifted seasonally or event-wise.

Evidence/Proxies: Sectional plans, notation of portable furniture and staging areas, and cross-section examples are documentary proxies indicating flexibility.

Assessment: The proposal scores high on nominal flexibility. Practical flexibility will depend on construction choices and maintenance regimes; lightweight, demountable fittings should be prioritized to preserve this capacity.

Figure 3.7 summarizes the landscape architecture design for the Libreville waterfront, "Bord de Mer," highlighting the transition from sea to urban infrastructure. The project includes a beach area, pedestrian promenade, and main boulevard, aiming to activate public spaces and improve water access through an "urban acupuncture" approach.

3.4.6 Analyzing Parameter VI—Predictability

Definition & rationale: Predictability in public-space design refers to users' ability to anticipate how spaces function and what behaviour is appropriate; this supports repeat use and lowers barriers to engagement (Norman, 2013).

Application to LibreLigne: The design creates clear nodes, visual markers and continuous circulation (pedestrian crossings, bike lanes), which improve legibility and therefore predictability of use. The themed piers create legible expectations (e.g., the Music Pier hosts performances).

Evidence/Proxies: Diagrams that show consistent signage, node identity and movement hierarchies, support legibility. However, because the project is not built at scale, there are no longitudinal observations of habitual routines that would confirm predictability in practice.

Fig. 3.8 View from one of the piers. *Source*: ecosistema urbana (2014)

Assessment: Design-led predictability is moderate to strong; empirical confirmation awaits implementation and observation of routine uses over time.

Figure 3.8 summarizes the bioclimatic strategy emphasizes outdoor sustainability, featuring a waterfront boulevard and unique landscape markers along the promenade.

3.4.7 Analyzing Parameter VII—Place Attachment (Proxy-Based Analysis)

Definition & rationale: Place attachment comprises emotional bonds, cognitive meanings and behavioral engagement with a place (Lewicka, 2011; Scannell & Gifford, 2010). Where direct surveys are unavailable, visual, programmatic and documentary proxies can indicate attachment (e.g., recurrent use, symbolic markers, and stewardship actions).

Emotional bonds (proxy): Project emphasis on local culture, themed piers and civic programming suggests potential for positive affective responses and local pride. Evidence to seek: photographic continuity of local events, media reports about community reactions, and documented stakeholder endorsements. Without implementation, emotional attachment remains a projected outcome rather than an observed state.

Cognitive meaning (proxy): The deliberate assignment of distinct themes to each jetée supports mnemonic mapping (residents can mentally locate "the Music Pier" or "the Culture Pier"), strengthening cognitive attachment. Proxies: interpretive materials, mapping of local narratives in the project documents, and app content plans that archive events and stories.

Behavioral engagement (proxy): The design's continuous promenade, bike lanes and program areas are likely to stimulate walking, cycling and repeated attendance at events—behavioral indicators of attachment. Proxies include documented programming schedules, site photographs showing repeated uses (if pilot events occurred), and design features that invite appropriation (benches, market spaces).

Assessment: LibreLigne's design is conducive to building place attachment through symbolic markers, programming and connective infrastructure. Empirical verification requires post-implementation observation; however, the proposal contains the typical design ingredients known to foster attachment (Lewicka, 2011; Scannell & Gifford, 2010).

3.4.8 Conclusion of Case Study I—LibreLigne

LibreLigne is a conceptually coherent application of urban-acupuncture thinking to a heritage waterfront: it concentrates identity and activity at a few precise nodes, embeds multifunctionality and flexibility in the built form, and explicitly links programming to cultural identity.

In terms of the six operational parameters, the design scores as follows: high on multi-functionality and flexibility; moderate–high on coping and adaptive capacity (design intent present); moderate on predictability; and promising on place attachment. The principal limitation is that LibreLigne remains a proposal: without implemented pilots, long-term evidence for everyday use, community uptake, and institutional transformation is absent. To move from design potential to verified outcomes, the project would require phased pilot implementations, monitoring of proxy indicators (event logs, photographic inventories, pedestrian counts), and policy instruments that secure local participation and affordability (Table 3.2).

3.5 Analyzing Case Study II—Kodak Passageway—Cairo, Egypt (2015)

Kodak Passageway was originally a narrow service alley connecting Adli Street and Abdel-Khalik Tharwat Street in downtown Cairo. Although informally used as a pedestrian shortcut during the 1990s, the passageway remained spatially neglected and socially underutilized.

Table 3.2 Summary of parameters of case study 1

Parameters	Description	Evidence
Coping capacity	The major purpose is to reconnect Libreville with the seaside and to extend the city's life to the waterfront in certain areas It provides public nodes for short-term social/economic activities	Design platforms for markets and events; conceptual evidence from design programming diagrams
Adaptive capacity	Urban regeneration operations tailored to the existing as well as the specific cultural, sociological and economic settings Modular design and thematic nodes enable programming adjustments	Modular jetée logic and continuous promenade allow adjustment over time
Transformative capacity	The project aims to integrate the concept of waterfront transformation into the context and community Potential to foster governance change if coupled with policy	Potential institutional shift if supported by governance; currently speculative
Multi-functionality	Combining, stretching, or contracting the intervention points results in a variety of profiles and interest regions Designed to host diverse uses across time	Explicitly programmed multi-use nodes across time and activities
Flexibility	The project is flexible because it gives the user multiple uses according to their needs Lightweight, reconfigurable interventions	Lightweight structures and varying pier profiles support reconfiguration
Predictability	The project lacks the capacity of predictability on the community Clear legibility but no long-term routine data	Legible nodes and circulation paths support user orientation
Place attachment	The mediating psychological construct (emotional, cognitive, behavioral bonds) that translates interventions into social effects	Symbolic nodes, and cultural programming and circulation foster attachment

A Kodak store, labs, garage and warehouse occupied the west side, and a branch of the once-popular Brazilian Coffee Stores chain was on the east side, and remains there today, although shuttered. Surrounded nearby by a number of emerging art, film and design spaces, the Kodak Passageway was identified by CLUSTER as an opportunity to create a pop-up gallery space, that might serve as a catalyst for urban redevelopment and revitalization of surrounding buildings, shops and passageways.

In 2015, CLUSTER (Cairo Lab for Urban Studies, Training, and Environmental Research) redesigned the passageway as a pedestrian-oriented public space through a small-scale urban acupuncture intervention. The project introduced improved paving, seating, art-deco-inspired lighting, and greenery, transforming the alley into a calm public corridor within a dense commercial district. The intervention aimed to enhance walkability, improve everyday urban experience, and reactivate social life without altering the historic fabric of the surrounding buildings. It aims to promote a more diverse, inclusive, and accessible public space in the heart of Cairo.

CLUSTER inaugurated a re-design of the public space of the passageway itself. Today, this 'Green Oasis' offers a landscaped, peaceful refuge featuring the revival of original 1930's art deco light fixtures and a custom designed walkway (Table 3.3).

Table 3.3 ID card for Kodak Passageway—Egypt

Kodak Passageway ID card	
See Fig. 3.9	
Project Name	Kodak Passageway
Location	Cairo, Egypt
Date of opening	January, 2015
Design and implementation	CLUSTER (Cairo Lab for Urban Studies, Training, and Environmental Research)
Civil engineering supervision	Eng. Atef Bassily

Fig. 3.9 Kodak Passageway, Egypt. Before-and-after images illustrating the transformation of a service alley into a pedestrian public space. *Source*: CLUSTER (2015)

3.5.1 Analyzing Parameter I—Coping Capacities

Definition & rationale: Coping capacity refers to the ability of individuals and communities to manage everyday stresses and disturbances through available spatial, social, and organizational resources (UNDRR, 2015).

Application to Kodak Passageway: The intervention directly addressed daily urban stressors in downtown Cairo, including pedestrian congestion, lack of resting spaces, and absence of greenery. By introducing seating, shading, and safe pedestrian circulation, Kodak Passageway provides a micro-scale refuge that supports short-term coping for workers, residents, and passersby. The improved lighting and clear visibility also enhance perceived safety, enabling sustained everyday use.

Evidence/proxies: Photographic documentation of pedestrian use, seating occupation, and informal pauses along the passageway serve as proxies indicating enhanced coping capacity. The transformation from a service alley into a usable public corridor suggests increased spatial support for everyday urban life.

Assessment: Kodak Passageway demonstrates strong coping capacity by mitigating immediate environmental stress and improving pedestrian comfort in a dense historic district.

Figure 3.10 shows the construction and upgrading phase of the passageway, showing surface improvements, lighting installation, and spatial reconfiguration aimed at improving pedestrian comfort and coping capacity in a dense historic district.

Fig. 3.10 KODAK Passageway during implementation. *Source*: CLUSTER (2015)

3.5.2 Analyzing Parameter II—Adaptive Capacities

Definition & rationale: Adaptive capacity is the ability of a system to adjust spatial and social practices in response to changing needs and conditions (UNDP, 2003).

Application to Kodak Passageway: The project followed an iterative process involving citizen workshops, a design competition, implementation, and post-implementation observation. This participatory and incremental approach allowed the design to respond to local needs and adapt over time. The flexible layout of seating and open circulation enables the passageway to accommodate varying flows, informal activities, and temporal changes throughout the day.

Evidence/proxies: Design process documentation, workshop records, and observational images of varied usage patterns indicate adaptive capacity embedded in both the process and the spatial outcome.

Assessment: Adaptive capacity is high, particularly due to the participatory design methodology and the open-ended nature of the spatial configuration.

Figure 3.11 shows the landscape and spatial design strategy for Kodak Passageway, illustrating seating placement, paving continuity, and circulation clarity that support adaptive use and flexible everyday appropriation.

Fig. 3.11 The design of the walkway landscape. *Source*: CLUSTER (2015)

3.5.3 Analyzing Parameter III—Transformative Capacities

Definition & rationale: Transformative capacity involves deeper changes in institutional practices, social norms, and perceptions of space (Walker et al., 2004).

Application to Kodak Passageway: By reclassifying a neglected service alley as a legitimate public space, the project challenges conventional hierarchies of urban value in downtown Cairo. CLUSTER's intervention demonstrates how informal and residual spaces can be reimagined as civic assets, potentially influencing municipal attitudes toward small-scale public space interventions.

Evidence/proxies: The project's documentation, dissemination through academic and professional platforms, and replication of similar alleyway interventions in Cairo act as indicators of transformative potential.

Assessment: While limited in physical scale, Kodak Passageway exhibits moderate transformative capacity by shifting perceptions of what constitutes meaningful public space in historic urban contexts.

Figure 3.12 shows the Conceptual diagram explaining the transformation of the alley into a pedestrian public corridor, highlighting spatial organization, seating zones, and lighting elements that reframe the passageway as a civic space.

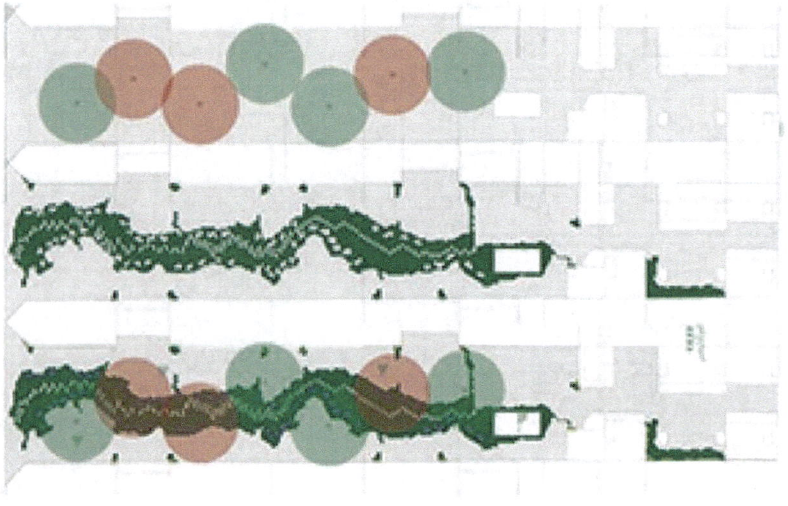

Fig. 3.12 Diagram of the design. *Source*: CLUSTER (2015)

3.5.4 Analyzing Parameter IV—Multi-Functionality

Definition & rationale: Multi-functionality refers to the integration of multiple uses within the same space across time (Brandt & Vejre, 2004). The project integrated the local economies into the design of the passageway.

Application to Kodak Passageway: The passageway accommodates movement, resting, informal social interaction, and small-scale economic activities linked to surrounding shops. Its design supports both transitory use and lingering, enabling coexistence of circulation and social functions.

Evidence/proxies: Observational photographs showing pedestrians walking, sitting, and engaging with adjacent storefronts demonstrate temporal and functional diversity.

Assessment: Kodak Passageway successfully achieves multi-functionality within a very limited spatial footprint.

Figure 3.13 represents perspective views showing the integration of movement, resting, and informal social interaction within a narrow spatial section, demonstrating the multi-functionality of the intervention.

Fig. 3.13 Perspective of the walkway. *Source*: CLUSTER (2015)

3.5.5 *Analyzing Parameter V—Flexibility*

Definition & rationale: Flexibility is the capacity of a space to accommodate changing uses with minimal physical alteration (Till & Schneider, 2005).

Application to Kodak Passageway: The open-ended design allows the passageway to respond to varying user demands throughout the day. Seating elements, open paving, and unobstructed circulation support spontaneous appropriation without fixed programming. The usage of the passageway changes with time according to the user needs and requirements.

Evidence/proxies: Variations in user behavior at different times, as documented in project photographs, serve as indicators of flexibility.

Assessment: The project demonstrates high flexibility, enabling continuous reinterpretation by users without structural modification.

Figure 3.14 shows the post-intervention condition of the passageway, illustrating increased pedestrian use, improved lighting, and enhanced spatial legibility that support flexibility and everyday social resilience.

Fig. 3.14 The Kodak Passageway after upgrading. *Source*: Abaza (2017)

3.5.6 Analyzing Parameter VI—Predictability

Definition & rationale: Predictability refers to users' ability to anticipate how a space functions, encouraging repeated use and social comfort (Norman, 2013).

Application to Kodak Passageway: Clear spatial cues—consistent lighting, defined walking paths, and visible seating—make the passageway legible and intuitive. Users can easily predict appropriate behaviors such as walking, resting, or brief social interaction.

Evidence/proxies: The orderly spatial layout and repeated patterns in user behavior suggest a high level of predictability.

Assessment: Predictability is strong, supporting routine use and reinforcing the passageway's role as a dependable urban connector.

Figure 3.15 shows nighttime view of the passageway highlighting lighting design and spatial clarity, reinforcing predictability, perceived safety, and routine use within the historic urban fabric.

Fig. 3.15 The Kodak Passageway at night. *Source*: CLUSTER (2015)

3.5.7 Analyzing Parameter VII—Place Attachment (Proxy-Based Analysis)

Definition & rationale: Place attachment encompasses emotional bonds, cognitive meaning, and behavioral engagement with place (Lewicka, 2011; Scannell & Gifford, 2010).

Emotional bonds (proxy): The calm atmosphere, lighting, and seating create conditions for positive affect and comfort, supporting emotional attachment. Continued everyday use by pedestrians and workers indicates affective acceptance.

Cognitive meaning (proxy): The passageway's identity as "Kodak Passageway" and its recognizable design elements contribute to its mental mapping as a distinct place rather than a residual alley.

Behavioral engagement (proxy): Recurrent walking, stopping, and informal social interaction indicate behavioral attachment and everyday appropriation.

Assessment: Kodak Passageway demonstrates a strong capacity to foster place attachment through everyday use, symbolic recognition, and behavioral engagement, despite its modest scale.

3.5.8 Conclusion of Case Study II

Kodak Passageway exemplifies how a small-scale urban acupuncture intervention can significantly enhance place attachment and social resilience in a historic urban environment. The project performs strongly across coping, adaptive, flexibility, predictability, and multi-functionality parameters, while also demonstrating meaningful—though modest –transformative potential. Its greatest strength lies in its ability to convert an overlooked residual space into a socially embedded public place that supports daily life, reinforces pedestrian networks, and restores a sense of belonging without triggering displacement or large-scale physical change (Table 3.4).

3.6 Analyzing Case Study III—Al Rusafa Historical City—Baghdad, Iraq (2020)

The historical center of Al-Rusafa is located on the eastern bank of the Tigris River and represents one of Baghdad's most significant urban cores at both local and regional levels. Al-Rusafa is vital at the local and regional levels, because it is related to many historical events and its content of a large number of traditional buildings, axes and squares. These historical areas display diverse architectural styles influenced

Table 3.4 Summary of parameters of case study 2

Parameters	Description	Evidence
Coping capacity	Kodak aims to deal with the problems of walkability and lack of greenery in the area	Improved walkability, seating, safety, and comfort
Adaptive capacity	The project started with citizen workshops, then design competition, implementation, and observation	Participatory design and open-ended spatial layout
Transformative capacity	CLUSTER analyzed the passageway and the problem with the lack of landscape and user-friendly design as the seating area	Reframing service alley as civic space
Multi-functionality	The project integrated the local economies into the design of the passageway	Movement, resting, and social interaction
Flexibility	The usage of the passageway changes with time according to the user needs and requirements	Informal and time-based appropriation
Predictability	The ability of a user to foresee which operations can be carried out next is one aspect of the predictability principle	Clear spatial cues and intuitive use
Place attachment	The mediating psychological construct (emotional, cognitive, behavioral bonds) that translates interventions into social effects	Strong daily engagement and symbolic recognition

by different historical phases, comprising mosques, schools, markets, and traditional neighborhoods. However, they face common issues such as deterioration, neglect, and a harmful shift from residential to industrial uses. The region features numerous historical buildings and public squares that represent the city's cultural identity, but unregulated developments and infrastructure decay have disturbed the area's spatial coherence and social vitality, resulting in various urban, social, economic, and environmental problems that contribute to significant urban decay.

The urban acupuncture strategy proposed by "Al-Hinkawi and Al-Saadi" (2020) identifies a network of critical nodes, landmarks, axes, and public spaces for targeted interventions. Rather than pursuing large-scale redevelopment, the project seeks to restore resilience by activating key points within the historic network, reinforcing connectivity, cultural continuity, and everyday social life (Table 3.5).

Table 3.5 ID card for Al Rusafa Historical City—Iraq

Kodak Passageway ID card	
See Fig. 3.16	
Project Name	Urban acupuncture in Al Rusafa
Location	Baghdad, Iraq
Date of opening	January, 2020
Designers	W. Sh. Al-Hinkawi and S. M. Al-Saadi
Urban strategy	Urban acupuncture

Fig. 3.16 Al Rusafa Historical City, Iraq. Urban network analysis illustrating key landmarks, axes, and intervention points within the historic core. *Source*: Al Hinkawi (2020)

3.6.1 Analyzing Parameter I—Coping Capacities

Definition & rationale: Coping capacity refers to the ability of people, institutions, and urban systems to manage disturbances, stress, and everyday challenges using existing resources and skills (UNDRR, 2015).

Application to Al-Rusafa: Al-Rusafa's historic core has experienced prolonged deterioration due to informal construction, incompatible land uses, and infrastructural neglect. These conditions have reduced the community's ability to cope with environmental stress, social fragmentation, and economic decline. The proposed urban acupuncture strategy responds by identifying weak points within the urban fabric and targeting them for rehabilitation, aiming to restore basic functionality, legibility, and social usability.

Evidence/proxies: Mapping of weak urban points, deteriorated public spaces, and disrupted pedestrian networks serves as a proxy for reduced coping capacity, while the proposed interventions indicate potential recovery mechanisms.

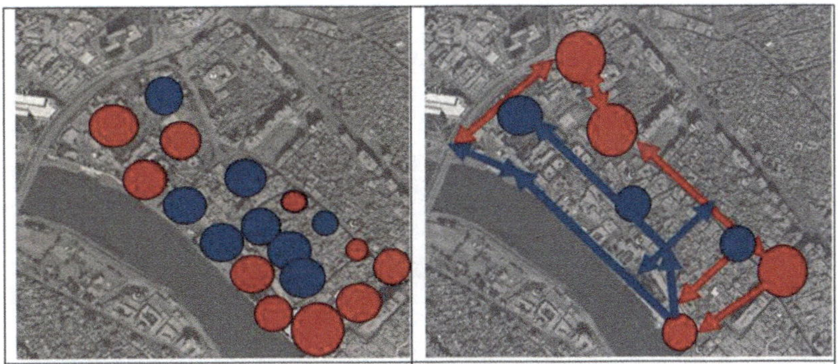

A. Identifying the hot and weak points landmarks and buildings | B. Identifying the hot and weak points in axes and nodes

Fig. 3.17 Mapping of degraded areas and spatial discontinuities targeted for intervention. *Source*: Al-Hinkawi and Al-Saadi (2020)

Assessment: The project addresses coping capacity at an urban-system level by stabilizing critical points within the historic network rather than focusing on isolated architectural objects.

Figure 3.17 shows the identification of weak urban points, deteriorated public spaces, and disrupted pedestrian networks within Al-Rusafa. The mapping serves as a proxy indicator for reduced coping capacity and informs the selection of targeted urban acupuncture interventions.

3.6.2 Analyzing Parameter II- Adaptive Capacities

Definition & rationale: Adaptive capacity is the ability of an urban system to adjust its spatial and social configurations in response to changing conditions (UNDP, 2003).

Application to Al-Rusafa: While the historic urban fabric has adapted informally to modern pressures, these adaptations often occurred at the expense of identity, spatial coherence, and social continuity. The proposed project reframes adaptation as a controlled and informed process, using urban acupuncture to guide future changes while respecting historical patterns. Strategic infill, reactivation of public spaces, and reinforcement of traditional axes enable adaptive reuse without erasing cultural memory.

Evidence/proxies: Proposed intervention points, adaptive reuse strategies, and network-based planning illustrate the project's capacity to redirect adaptation toward resilience rather than degradation.

Assessment: Adaptive capacity is moderate to high, as the project explicitly responds to past maladaptive transformations and proposes a more resilient future trajectory.

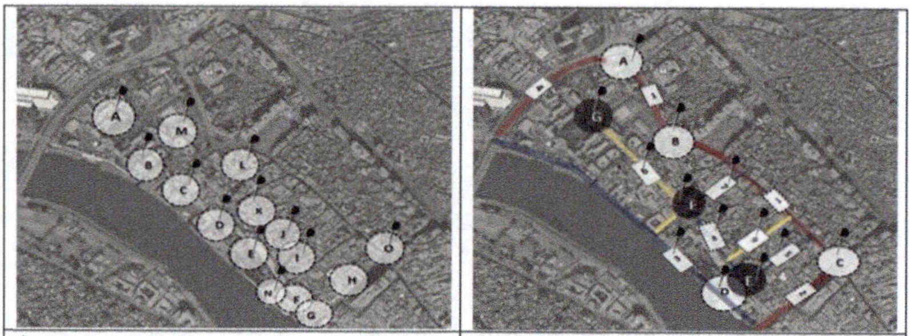

A. The proposed acupuncture points for interventions (distinctive landmarks and buildings)

B. The proposed acupuncture points for interventions (distinctive axes and nodes)

Fig. 3.18 Strategic nodes and corridors selected to guide adaptive transformation. *Source*: Al-Hinkawi and Al-Saadi (2020)

Figure 3.18 displays two diagrams (A and B) illustrating the concept of "urban acupuncture" applied to the historic center of Rusafa, Baghdad. Diagram A shows proposed intervention points based on distinctive landmarks and buildings. Diagram B shows proposed intervention points based on distinctive axes and nodes, highlighting movement and connections.

3.6.3 Analyzing Parameter III—Transformative Capacities

Definition & rationale: Transformative capacity involves deeper structural change in values, governance, and spatial practices, enabling long-term resilience (Walker et al., 2004).

Application to Al-Rusafa: The project demonstrates strong transformative intent by redefining how historic cities can be revitalized. By prioritizing landmarks, axes, and social nodes as catalysts, the intervention challenges conventional large-scale redevelopment models and promotes a culturally grounded, community-oriented approach. This shift encourages new institutional and planning attitudes toward heritage-led resilience.

Evidence/proxies: The selection of culturally significant nodes and the emphasis on social regeneration indicate transformative capacity beyond physical repair.

Assessment: Transformative capacity is high, particularly in its potential to influence heritage management and urban policy in post-conflict and rapidly changing cities.

Figures 3.19 represents a diagram showing proposed nodes, axes, and corridors selected to guide adaptive transformation within the historic urban fabric. The figure illustrates how controlled, heritage-sensitive interventions redirect adaptation toward resilience rather than informal degradation.

Fig. 3.19 Suggested acupuncture for the study area (for green areas). *Source*: Al-Hinkawi amd Al-Saadi (2020)

Figure 3.20 shows the final stage where the city is reconnected and restructured according to the proposed procedures and interventions. It is based on urban network analysis illustrating key landmarks, axes, public spaces, and proposed intervention points within the historic core. The figure highlights the spatial logic of the urban acupuncture strategy, emphasizing connectivity, cultural continuity, and node-based resilience within the heritage fabric.

3.6.4 Analyzing Parameter IV- Multi-Functionality

Definition & rationale: Multi-functionality refers to the integration of multiple social, cultural, and economic functions within shared spaces over time (Brandt & Vejre, 2004).

Application to Al-Rusafa: Unlike the previous case studies, this project operates at an urban-network scale. By reconnecting landmarks, public spaces, commercial streets, and religious sites, the intervention supports overlapping functions such as mobility, commerce, social interaction, and cultural expression. Nodes serve as multifunctional anchors rather than single-purpose destinations.

Fig. 3.20 Rebuilding the final urban network for the study area. *Source*: Al-Hinkawi and Al-Saadi (2020)

Evidence/proxies: Urban network diagrams and proposed functional overlaps demonstrate the project's high multi-functional capacity.

Assessment: Multi-functionality is very high due to the project's scale and integrative spatial logic.

3.6.5 Analyzing Parameter V—Flexibility

Definition & rationale: Flexibility refers to the capacity of a system to accommodate change without losing its core identity (Till & Schneider, 2005).

Application to Al-Rusafa: The project's flexibility is grounded in extensive community and contextual analysis conducted prior to intervention. Rather than prescribing fixed uses, the strategy allows different nodes to evolve according to local needs while maintaining historical continuity. This enables gradual transformation and user-driven appropriation over time.

Evidence/proxies: Phased intervention strategies and adaptable node-based planning serve as indicators of flexibility.

Assessment: Flexibility is high, particularly due to the project's non-deterministic and incremental nature.

3.6.6 Analyzing Parameter VI—Predictability

Definition & rationale: Predictability supports user confidence by making spatial behavior and movement legible and intuitive (Norman, 2013).

Application to Al-Rusafa: One of the major challenges in the historic core is the loss of spatial clarity caused by haphazard additions and disrupted visual axes. The proposed intervention restores predictability by reinforcing traditional paths, landmarks, and spatial hierarchies. Clear node-based organization allows users to anticipate movement, gathering points, and functional transitions.

Evidence/proxies: The re-established network of axes and landmarks functions as a cognitive map for users.

Assessment: Predictability is significantly strengthened through the restoration of historic spatial logic.

3.6.7 **Analyzing Parameter VII—Place Attachment (Proxy-Based Analysis)**

Definition & rationale: Place attachment encompasses emotional bonds, cognitive meaning, and behavioral engagement with place (Lewicka, 2011; Scannell & Gifford, 2010).

Emotional attachment (proxy): Revitalizing culturally meaningful landmarks and public spaces fosters emotional reconnection with place, particularly in a post-conflict context.

Cognitive attachment (proxy): Reinforcing historic axes and recognizable landmarks strengthens collective memory and mental mapping.

Behavioral attachment (proxy): Encouraging everyday use of restored nodes and paths supports routine engagement and social appropriation.

Assessment: The project demonstrates strong potential to rebuild place attachment by aligning physical interventions with cultural memory and social practices.

3.6.8 Conclusion of Case Study III

The Al-Rusafa urban acupuncture project represents a strategic, heritage-sensitive approach to resilience in a historic Middle Eastern city. By operating at the scale of the urban network, the intervention exhibits strong transformative, multi-functional, and flexible capacities while addressing long-standing coping and predictability challenges. The project's emphasis on landmarks, nodes, and cultural continuity positions it as a powerful model for rebuilding place attachment and social resilience in historic urban environments affected by long-term degradation and instability (Table 3.6).

Table 3.6 Summary of parameters of case study 3

Parameters	Description	Evidence
Coping capacity	Over time, the haphazard additions that have compromised their aesthetic integrity have accumulated urban issues of all stripes—urban, social, economic, and environmental—leading to a serious decline in metropolitan areas	Targeting weak urban points and degraded public spaces
Adaptive capacity	The urban and social fabric of the city have adapted to the new infill projects that ruined its identity and hence there were damages on the urban and social levels	Redirecting informal adaptation toward resilience
Transformative capacity	The landmarks, buildings, axes and nodes selected for interventions are base points for transformation that leads to resilience	Heritage-led, node-based urban regeneration
Multi-functionality	The project has a different level than the previous one. As an urban level project, connecting nodes and buildings and landmarks, it has a high multi-functional capacity	Network-scale integration of social and cultural uses
Flexibility	The project's flexibility is based on the community studies done before the implementation, and according to the definition of flexibility, the project aims to make the community of AL-Rusafa resilient	Incremental, community-informed interventions
Predictability	The ability of a user to foresee which operations can be carried out next is one aspect of the predictability principle	Restoration of historic axes and landmarks
Place attachment	The mediating psychological construct (emotional, cognitive, behavioral bonds) that translates interventions into social effects	Reinforcement of cultural memory and everyday use

3.7 Comparative Analysis Matrix of the Case Studies

See Table 3.7.

Table 3.7 Comparative analysis of case studies

Parameter	LibreLigne (Libreville, 2014)	Kodak Passageway (Cairo, 2015)	Al-Rusafa Historical Center (Baghdad, 2020)
Coping capacity	Reconnects the city with the seaside and extends urban life to the waterfront	Addresses walkability problems and lack of green public space	Responds to accumulated urban, social, economic, and environmental decline caused by unplanned additions
Adaptive capacity	Urban regeneration tailored to cultural, sociological, and economic contexts	Citizen workshops followed by a design competition, implementation, and observation	Urban and social fabric adapted –often negatively—to infill developments that damaged identity
Transformative capacity	Integrates waterfront transformation within the community context	Reconfigures the alley through improved seating and landscape design	Landmarks, axes, and nodes selected as transformation points leading to resilience
Multi-functionality	Variable intervention points generate diverse spatial profiles and activities	Integrates local economic activities into the passageway design	Urban-scale connectivity enables high multifunctional capacity
Flexibility	Allows multiple uses according to user needs	Usage varies temporally based on user demands	Community-based planning supports adaptive flexibility and resilience
Predictability	Moderate–strong; long-term routines yet to be empirically observed	Strong predictability reinforcing daily use and safety	Predictability improved through restoration of historic spatial logic
Place attachment—(emotional)	Emerging affective bonds through cultural programming and themed spaces	Growing emotional attachment as a reclaimed public oasis	Deep-rooted emotional bonds tied to history and rituals
Place attachment—(cognitive)	Mnemonic landmarks and narratives structure meaning and memory	Art-Deco restoration reinforces historic identity	Very strong collective memory and identity embedded in heritage fabric
Place attachment—(behavioral)	Encourages recurring practices; post-occupancy data needed	Increased footfall, events, and informal commerce observed	Sustained everyday use and community stewardship practices

3.8 Conclusion of the Analytical Part

This chapter has provided a structured and comparative analysis of three urban acupuncture interventions in heritage contexts—LibreLigne in Libreville, Kodak Passageway in Cairo, and Al-Rusafa Historical City in Baghdad—using a unified analytical framework grounded in social resilience theory and place attachment psychology. By systematically applying the same parameters across all cases, the chapter ensured analytical consistency while allowing contextual differences in scale, governance, and socio-cultural conditions to emerge clearly.

The analysis demonstrated that urban acupuncture functions not merely as a spatial design strategy, but as a socio-psychological catalyst, capable of mediating the impacts of structural pressures—such as gentrification, and socio-spatial disruption—through the activation of place attachment. Emotional, cognitive, and behavioral bonds were shown to play a critical role in translating spatial interventions into resilient social outcomes, whether through everyday use, collective memory, or informal stewardship.

Across the three cases, different resilience profiles became evident. **LibreLigne** revealed the strengths and limitations of conceptually robust but largely speculative projects, highlighting high adaptive and multi-functional potential that remains contingent on implementation and governance mechanisms. **Kodak Passageway** illustrated how small-scale, implemented interventions can rapidly enhance coping and adaptive capacities through improved predictability, walkability, and recurrent use. **Al-Rusafa**, operating at an urban-network scale, demonstrated strong transformative capacity rooted in deep historical and cognitive place attachment, while simultaneously exposing the constraints imposed by accumulated structural, institutional, and spatial challenges.

Throughout the three cases, urban acupuncture strategies vary by scale and maturity: LibreLigne is a robust design concept with high potential for multifunctionality and adaptability but lacks longitudinal evidence; Kodak Passageway is an implemented micro-intervention demonstrating tangible gains in predictability, use and behavioral attachment; Al Rusafa represents urban-scale acupuncture with strong transformative intent and deep pre-existing place attachment, but faces structural constraints from prior unsympathetic infill. Overall, interventions that combine design sensitivity to heritage with community participation show the strongest immediate gains in behavioral engagement and predictability, while larger, transformative projects require enabling policy to convert potential into sustained social resilience (Ecosistema Urbano, 2014; CLUSTER, 2015; Al-Hinkawi & Al-Saadi, 2020).

Taken together, the case studies confirm that the effectiveness of urban acupuncture is not determined by scale alone, but by the degree to which design strategies are aligned with existing place attachments and supported by enabling social and institutional conditions. Interventions that reinforce meaning, encourage repeated everyday practices, and respect inherited socio-spatial patterns are more likely to contribute to sustained social resilience than purely aesthetic or isolated physical upgrades.

While the analysis relied primarily on qualitative evidence and design-based proxies rather than longitudinal empirical data, this approach remains appropriate for comparative heritage research and offers a solid interpretive foundation. Importantly, the chapter establishes a clear analytical bridge between theory and practice, demonstrating how abstract concepts—such as place attachment and social resilience—can be operationalized within urban design assessment.

This concluding synthesis thus sets the groundwork for the final discussion and conclusion of the study. The final conclusion will integrate these analytical insights at a broader theoretical and urban scale, reflect on their implications for heritage contexts facing contemporary pressures, and articulate recommendations for designers, planners, and policymakers seeking to employ urban acupuncture as a resilience-oriented strategy.

References

Abaza, M. (2017). Cairo: Restoration? And the limits of street politics. *Space and Culture, 20*(2), 170–190.

Al-Hinkawi, W. S., & Al-Saadi, S. M. (2020). Urban acupuncture, a strategy for development: case study of Al-Rusafa, Baghdad. In *IOP Conference series: Materials science and engineering* (Vol. 881, No. 1, p. 012002). IOP Publishing.

Brandt, J., & Vejre, H. (2004). Multifunctional landscapes—advances in ecological planning. Landscape and Urban Planning.

Ecosistema Urbano. (2014). *LibreLigne project materials and diagrams*. (Project materials). Retrieved from https://ecosistemaurbano.com/

Kodak Passageway—Cairo Urban Initiatives Platform. CLUSTER (Cairo Lab for Urban Studies, Training, and Environmental Research). Retrieved from https://passageways.clustermappinginitiative.org/en/pilot-project/kodak-passageway

Lewicka, M. (2011). Place attachment: How far have we come in the last 40 years? *Journal of Environmental Psychology, 31*(3), 207–230. https://doi.org/10.1016/j.jenvp.2010.10.001

Norman, D. A. (2013). *The design of everyday things*. Basic Books.

Scannell, L., & Gifford, R. (2010). Defining place attachment: A tripartite organizing framework. *Journal of Environmental Psychology, 30*(1), 1–10. https://doi.org/10.1016/j.jenvp.2009.09.006

Till, J., & Schneider, T. (2005). Flexible housing: The means to the end. *ARQ: Architectural Research Quarterly, 9*(3–4), 287–296.

UNDP. (2003). *Adaptation policy frameworks for climate change*. United Nations Development Programme.

UNDRR. (2015). *Sendai framework for disaster risk reduction 2015–2030*. United Nations Office for Disaster Risk Reduction.

Walker, B., Holling, C. S., Carpenter, S. R., & Kinzig, A. (2004). Resilience, adaptability and transformability in social–ecological systems. *Ecology and Society, 9*(2), 5.

Chapter 4
Discussion and Conclusion

4.1 Discussion

Building on the comparative insights derived from the three case studies, this book moves beyond individual project analysis to critically interpret the findings within a broader theoretical and urban discourse. While we had demonstrated how urban acupuncture interventions operate across different socio-spatial and heritage contexts, the present discussion seeks to synthesize these observations, identify cross-cutting patterns, and evaluate their implications for understanding social resilience in cities under structural pressure. By reflecting on the interplay between gentrification dynamics, spatial design strategies, and place attachment mechanisms, this chapter positions the case studies as empirical lenses through which the study's conceptual framework is tested, refined, and expanded. By doing so, it not only addresses the practical outcomes of these interventions, but also explains why certain configurations are more effective in sustaining community continuity, adaptive capacity, and long-term resilience in heritage environments.

1. Reframing Urban Acupuncture through Social Resilience

This book set out to investigate how small-scale, targeted urban interventions can operate as catalysts for social resilience in historic urban environments subjected to gentrification pressures. Through the analytical framework developed and empirically tested across three contrasting case studiesLibreLigne in Libreville, Kodak Passageway in Cairo, and Al-Rusafa Historical City in Baghdadthe research demonstrates that urban acupuncture should be understood as a psychosocial and institutional process integrated into larger socio-spatial dynamics rather than just as a design strategy.

Across all cases, urban acupuncture interventions functioned as strategic disruptions within broader urban systems, revitalizing neglected areas, reorganizing daily routines, and renegotiating human-place interactions and connections. However, the extent to which these interventions contributed to social resilience depended not on

B. I. Farahat and H. O. Alaeddine, *Urban Acupuncture for Place Attachment and Social Resilience in Heritage Contexts*, Creativity, Heritage and the City 10,
https://doi.org/10.1007/978-981-95-8970-8_4

their physical form alone, but on how effectively they mediated place attachment, responded to gentrification pressures, and aligned with existing social and cultural structures.

2. Gentrification as a Differentiated Structural Pressure

The comparative analysis reveals that gentrification operates differently across contexts, yet consistently acts as a structural condition shaping the stakes of intervention. In Libreville, gentrification pressure emerged primarily through speculative waterfront redevelopment and tourism-oriented visions; in Cairo, it manifested as micro-scale neglect, informal appropriation, and socio-economic exclusion; while in Baghdad, it appeared through unregulated infill, heritage erosion, and post-conflict urban fragmentation.

Rather than treating gentrification as a solely negative phenomenon, the findings suggest that its impacts are mediated by governance structures, cultural continuity, and community agency. Urban acupuncture proved most effective where it acknowledged gentrification pressures explicitly –by either redirecting development logics, re-centering local practices, or reinforcing existing social networksrather than attempting to neutralize them through design alone.

3. Place Attachment as the Critical Mediating Mechanism

One of the central contributions of this research lies in demonstrating that place attachment functions as the key psychological mediator between spatial intervention and social resilience. The emotional, cognitive, and behavioral dimensions of attachment provided a lens through which the social effectiveness of each project could be evaluated beyond aesthetic or functional success.

LibreLigne, although largely conceptual, showed strong potential for cognitive attachment through narrative construction, thematic nodes, and symbolic waterfront reconnection. The Kodak Passageway case illustrated how even modest spatial transformations can significantly strengthen emotional and behavioral attachment through safety, legibility, and everyday use. Al-Rusafa, by contrast, highlighted the resilience of pre-existing place attachment rooted in historical continuity, ritualized use, and collective memorydemonstrating that in some contexts, the role of urban acupuncture is not to create attachment, but to stabilize and protect it.

These findings demonstrate that urban acupuncture runs the potential of being superficial or fleeting if place attachment is not explored. Interventions enhanced stewardship, meaning-making, and long-term social investment in addition to use and appropriation when attachment was activated or reinforced.

4. Social Resilience as a Multi-Temporal Process

The operationalization of social resilience through coping, adaptive, and transformative capacities enabled a nuanced reading of how communities respond to spatial change over time. Coping capacities were most visible in short-term improvements to accessibility, safety, and daily stress reduction; adaptive capacities emerged through flexible programming, participatory processes, and iterative design; while transformative capacities depended heavily on institutional alignment and long-term governance frameworks.

The cases reveal that urban acupuncture most reliably strengthens coping and adaptive capacities, while transformative outcomes remain contingent and uneven. Transformative change was strongest in Al-Rusafa at the conceptual level, where interventions targeted structural nodes and urban hierarchies, yet weakest in practice due to institutional fragility. This underscores a critical insight: design-led transformation requires institutional resonance to become socially durable.

5. Methodological Reflections: The Value of Proxy Indicators

Given the limited availability of longitudinal social data and direct surveys in many heritage contexts, this research adopted proxy indicators such as spatial use patterns, programming diversity, symbolic elements, and documented community practices, to operationalize abstract psychological and social concepts. The analysis demonstrates that, when theoretically grounded and triangulated, such proxies offer a valid and rigorous means of evaluating social outcomes in urban research.

Rather than weakening the analysis, the use of proxies allowed for comparative consistency across cases and facilitated the translation of complex conceptssuch as place attachment and resilienceinto observable spatial and social phenomena. This methodological approach is particularly valuable in fragile, post-conflict, or informally governed contexts, where conventional data collection is often infeasible.

6. Synthesis and Final Reflections

Taken together, the findings of this book affirm that urban acupuncture is most effective when it operates at the intersection of space, psychology, and governance. Its capacity to counteract the destabilizing effects of gentrification does not lie in scale or spectacle, but in its ability to reinforce meaningful relationships between people and place, support everyday practices, and open pathways for adaptive and transformative change.

By integrating architectural psychology, place attachment theory, and social resilience into a unified analytical framework, this research contributes a multidimensional understanding of urban intervention in historic environments. It positions urban acupuncture not as an isolated design tool, but as a strategic framework for socially attuned urban regeneration, capable of sustaining heritage, community identity, and resilience in times of profound urban change.

4.2 Conclusion

In conclusion, this book shows that urban acupuncturewhen intentionally aligned with existing place attachmentcan meaningfully strengthen social resilience in historic urban environments: small, well-placed interventions restore everyday use, reinforce memory and meaning, and catalyze coping and adaptive responses, while larger networked actions hold transformative potential when paired with enabling governance. Effectiveness depends less on scale than on cultural fit, participatory processes, and policy support: design without stewardship, inclusion and monitoring risks short-lived gains or unintended displacement. Methodologically, the proxy-based approach offered a rigorous, comparable reading across diverse contexts, though empirical post-occupancy monitoring remains necessary to confirm long-term social outcomes. With these conclusions, the study now turns to concrete, design-oriented recommendations aimed at converting the documented potential of urban acupuncture into durable, inclusive resilience in heritage cities.

This book argues that cities do not only fracture through physical displacement, but through the quiet erosion of belonging. In historic neighborhoods under gentrification pressure, urban acupuncture emerges not as a cosmetic intervention, but as a precise act of careone that reconnects people to place, restores everyday meaning, and enables communities to endure change without losing themselves. By positioning place attachment as the psychological hinge between spatial intervention and social resilience, this research reframes small-scale design as a catalyst for long-term continuity in heritage environments.

4.3 Design-Oriented Recommendations

1. Design for small and strategic "nodes": Urban acupuncture works by concentrating effect where it amplifies local social life.
2. Embed place-attachment triggers into every intervention: Emotional, cognitive and behavioral bonds transform short-term activation into long-term stewardship.
3. Prioritize multi-functionality by design: Multi-functional places increase daily activation and economic resilience.
4. Design with stewardship in mind: Stewardship sustains resilience beyond project lifetimes.
5. Use digital storytelling and light monitoring to extend cognitive attachment: Digital traces broaden memory and connect Diasporas.

4.4 Book Summary

This book investigates the role of urban acupuncture as a design-led strategy for reinforcing place attachment and fostering social resilience within heritage contexts affected by gentrification and tourism pressures. Bridging architectural theory, urban design, and environmental psychology, the study advances an interdisciplinary framework that positions the human–place relationship at the core of heritage resilience.

Grounded in contemporary scientific theory, the book conceptualizes gentrification as a structural pressure that disrupts socio-spatial continuity, place attachment as a psychological mediator vulnerable to erosion, social resilience as a systemic outcome reflecting community capacity to cope, adapt, and transform, and urban acupuncture as an intentional, small-scale spatial intervention capable of repairing these disruptions. This conceptual apparatus is operationalized through clearly defined analytical parameters and proxy indicators, allowing for rigorous qualitative evaluation in contexts where direct social surveying is constrained.

Empirically, the book examines three international case studies: LibreLigne in Libreville, Kodak Passageway in Cairo, and Al-Rusafa Historical City in Baghdad, selected for their diversity in scale, cultural context, and implementation stage. Through comparative qualitative analysis, the study demonstrates how targeted micro-interventions can re-activate heritage spaces, strengthen emotional, cognitive, and behavioral dimensions of place attachment, and enhance social resilience outcomes across different urban conditions.

The findings reveal that design interventions aligned with local identity, everyday practices, and community participation are more effective in sustaining heritage environments than large-scale, top-down redevelopment approaches. The book argues that socially resilient heritage is not achieved through preservation alone, but through the continuous cultivation of belonging, meaning, and collective agency embedded in spatial practice.

By reframing urban acupuncture as a psychosocial design tool, this book contributes a novel methodological and theoretical lens to architectural and urban scholarship, offering architects, planners, and policymakers actionable insights for managing change in heritage environments while safeguarding social continuity, cultural identity and sense of belonging.

Bibliography

Holling, C. S., & Gunderson, L. H. (Eds.). (2012). *Panarchy: understanding transformations in human and natural systems*. Princeton University Press.

B. I. Farahat and H. O. Alaeddine, *Urban Acupuncture for Place Attachment and Social Resilience in Heritage Contexts*, Creativity, Heritage and the City 10,
https://doi.org/10.1007/978-981-95-8970-8

Index

B. I. Farahat and H. O. Alaeddine, *Urban Acupuncture for Place Attachment and Social Resilience in Heritage Contexts*, Creativity, Heritage and the City 10,
https://doi.org/10.1007/978-981-95-8970-8

GPSR Compliance

The European Union's (EU) General Product Safety Regulation (GPSR) is a set of rules that requires consumer products to be safe and our obligations to ensure this.

If you have any concerns about our products, you can contact us on ProductSafety@springernature.com

In case Publisher is established outside the EU, the EU authorized representative is:

Springer Nature Customer Service Center GmbH
Europaplatz 3
69115 Heidelberg, Germany

Batch number: 10837190

Printed by Printforce, the Netherlands